THE EVOLUTIONARY INTERACTION OF ANIMALS AND PLANTS

THE EVOLUTIONARY INTERACTION OF ANIMALS AND PLANTS

PROCEEDINGS OF
A ROYAL SOCIETY DISCUSSION MEETING
HELD ON 27 AND 28 FEBRUARY 1991

ORGANIZED AND EDITED BY
W. G. CHALONER, J. L. HARPER AND
J. H. LAWTON

LONDON
THE ROYAL SOCIETY
1991

Printed in Great Britain for the Royal Society
by the
University Press, Cambridge

First published in *Philosophical Transactions of the Royal Society of London.*
series B, volume 333 (no. 1267), pages 175–305

The text paper used in this publication is alkaline sized with a coating which is predominantly calcium carbonate. The resultant surface pH is in excess of 7.5, which gives maximum practical permanence.

British Library Cataloguing in Publication Data

A CIP catalogue record for this book is available from the British Library.

ISBN 0 85403 443 9

Published by the Royal Society
6 Carlton House Terrace, London SW1Y 5AG

PREFACE

The interaction between plants and animals has been a major factor in the evolution of both groups throughout the history of life on Earth. These interactions have ranged from casual herbivore–plant encounters to the intimate symbioses of photosynthetic microbes within the bodies of aquatic invertebrates, the finely tuned specialization of flower structure and pollination vector, and the chemical (and structural) warfare between plants and the animals that eat them, However, how far is it true that the plant kingdom is what the animal kingdom made it, and vice versa? This fundamental question and the rich details that permeate many aspects of the plant–animal interplay form the subject of this volume. Palaeobiologists, biochemists, cell biologists and a wide group of scientists working on whole-animal and whole-plant biology join forces to share their very different perspectives on this most challenging aspect of the evolutionary story. In thinking about who to invite to the meeting, we tried to avoid more familiar topics and to concentrate, instead, on those aspects of animal–plant interactions that have been less well studied or which seemed ripe for new syntheses or new interpretations. Our colleagues rose to the unconventional challenges we set them with great enthusiasm and skill.

A strangely assorted body of fossil evidence contributes to our understanding of evolutionary interactions of animals and plants. The gut contents of fossil animals, their fossilized droppings, bites out of leaves, the first appearance of floral nectaries and the pattern of wear on fossil mammal teeth are just some of the signals from the past that give a time dimension to the story. Throughout the history of their co-existence, the two groups have evidently influenced one another at various levels. An endosymbiosis of cyanobacterial cells with a eukaryotic cell brought the chloroplast to the plant kingdom. Why have so few animal groups adopted this admirable arrangement, to combine photosynthesis with holozoic nutrition? We have a scattering of aquatic invertebrates with photosynthetic symbionts, but why no green vertebrates? Or consider a different, but no less intriguing problem. A great variety of structures have evolved that allow animals to eat plants and deal with the way in which plants put their most nutritious materials inside cellulose boxes. Yet vertebrates and virtually all of the insects, the two major assailants of the plant world, have not evolved the capacity to digest cellulose, relying instead on microbes in their guts to do it for them. Why?

These and many other aspects of the plant–animal love–hate relationship are explored by the contributors to this volume. We hope that it not only summarizes areas of understanding, but also that it will stimulate others to seek new approaches to this most diverse field of evolutionary study.

We than most warmly all the contributors to this meeting, both the speakers and those who contributed to the discussion. Particularly, we are grateful for the prompt submission of manuscripts and return of proofs which has made it possible to keep to our tight publication schedule. We also record our grateful thanks to the Royal Society for organizing and hosting this meeting, and handling its publication. We thank particularly Christine Johnson, who worked heroically to keep us all on schedule, and Funda Suleyman and Simon Gribbin for their hard work in seeing the volume successfully to press.

August 1991

W. G. Chaloner
J. L. Harper
J. H. Lawton

CONTENTS

Fossil evidence for plant–arthropod interactions in the Palaeozoic and Mesozoic

W. G. CHALONER[1], A. C. SCOTT[2] AND J. STEPHENSON[2]

Biology[1] and Geology[2] Departments, Royal Holloway and Bedford New College, University of London, Egham Hill, Egham, Surrey TW20 0EX, U.K.

SUMMARY

Some of the earliest Devonian fossils of vascular plants show lesions that may be attributed to plant feeding activity by animals. This is the beginning of a more or less continuous fossil record of plant–animal interactions which extends from the Devonian to the present day. An important feature of pre-Cretaceous material is the evidence from coprolites and gut-contents of spore eating by arthropods. Experiments with living arthropods, of groups represented in the Palaeozoic, show that viable spores can survive passage through the gut in significant numbers. Spore eating could clearly have had a dispersal role of value to the plant, as well as its evident benefit as a source of nutrition for the animal involved. Evidence of wood boring and leaf eating extends from the late Carboniferous onwards. It appears that 'continuous marginal' leaf-feeding preceded 'interrupted marginal' feeding, and that this was in turn followed by 'non-marginal' leaf feeding. The latter first appeared in Cretaceous angiosperms. Some diversity of leaf miners and leaf galls are also represented in Cretaceous angiosperm leaf fossils.

1. INTRODUCTION

The wide range of plant–insect interactions in modern ecosystems has received a great deal of attention from biologists, not least because many of them have enormous economic significance. This is particularly so in the case of insect predation of crops of all kinds, both in the field and storage, and also in the insect pollination of angiosperms, with the commercial activities associated with fruit growing and honey production. In this review we consider some of the fossil evidence of how, and more particularly when, these relationships first appeared. Even dealing with living relationships, available for experimental study, details of the degree of mutuality of the relation are often open to debate. In fossil material, the exact nature of the relationship is inevitably less secure, and the point at which we can claim evidence of coevolution is blurred. We are confined to those few cases where there is tangible physical evidence in either the plant or animal remains, of the kinds of interaction for which we can see living analogues.

In the following review, we consider the evidence in a time sequence, starting with the earliest appearance of land-adapted arthropods and plants and their relationships in the terrestrial ecosystem. We concentrate on the earlier phase of this story, as the angiosperm–insect relationship in the context of biotic pollination (with a solely Cretaceous–Tertiary record) is dealt with in another contribution to this symposium (Crepet *et al.*). However, we deal briefly with the fossil evidence of other aspects of angiosperm–insect interaction, in the form of leaf feeding, leaf miners and gall formation.

2. PRE-DEVONIAN PLANT–ANIMAL INTERACTIONS

Plant fossils showing the characteristic features associated with terrestrial adaptation (stomata, a cuticle, xylem and wind-dispersed spores) do not appear until late in Silurian time. Certain of these attributes appear earlier in the Silurian and even late Ordovician, in the form of isolated spores and tissue fragments, but such evidence leaves open the real nature of the plants that produced them (Gray 1985; Edwards & Burgess 1990; Chaloner 1988). Plants attributed to the genus *Cooksonia* occur from latest Silurian rocks in Wales and the Welsh Border, and were probably inhabiting fluviatile areas close to a shoreline, or on tidal flats (Edwards *et al.* 1986). These were simple undifferentiated vascular plants with sporangia borne as swellings at the ends of the dichotomizing axes. Similar plants, but lacking secure evidence of vascular tissue and the other terrestrial adaptations cited above, occur in rocks as old as the mid-Silurian, and are sometimes referred to as 'rhyniophytoid' plants, by analogy with the securely tracheophyte genus *Rhynia* of the Early Devonian (Edwards & Feehan 1980). Rather more complex vascular plants, with clear stem–leaf (microphyll) differentiation are reported from the mid-Silurian of Australia, but their age relations with fossils from the northern continents remains controversial (Chaloner 1988; Edwards & Burgess 1990; Hueber 1983).

The earliest record of a diverse assemblage of terrestrial arthropods comes from latest Silurian (Pridolian) rocks from the Welsh Border (Jeram *et al.* 1990; Rolfe 1990), and includes two centipedes and a trigonotarbid arachnid. Although by analogy with their nearest living relatives these are likely to have been zoophagous forms, it is significant that they occur in association with fossils of the land plant *Cooksonia*. The shelter and humid environment produced by a sward of even quite small terrestrial plants, which were also offering a source of primary productivity, must have been one of the earliest and most critical interactions between plants and arthropods, making possible the land migration of the latter group.

Before looking at the subsequent course of this interaction, it must be acknowledged that there are various records of putative terrestrial life pre-dating this late Silurian plant–arthropod association. Rolfe (1985 *a*) interprets the late Silurian record of myriapods as representing the earliest occurrence of what were probably detritivores. Much earlier, we have record of fossil droppings (coprolites) from the Precambrian (Robbins *et al.* 1985) but these are inevitably of uncertain origin, and possibly are not even from a terrestrial animal. Mid-Silurian coprolites containing fungal material are described by Sherwood-Pike & Gray (1985), and these they attribute to fungivorous arthropods. However, it is possible in this case that the fungal material might have grown within the droppings, rather than having been the foodstuff for the arthropods. Even so, they are one of the earliest records of coprolites probably attributable to arthropods in a terrestrial setting.

Equally controversial are the burrows in a fossil soil profile from the Ordovician of Pennsylvania, described by Retallack & Feakes (1987). If they truly represent burrows made by sizeable organisms within a soil profile, as Retallack & Feakes suggest (they are from 3–16 mm in diameter and up to half a metre in depth), then we must assume that there was some form of primary production occurring in the terrestrial ecosystem of which they were a part. To that extent they represent evidence for at least an energy-flow interaction between elements of land-adapted fauna and some terrestrial autotrophs. The latter might have been simple algae living on a moist soil surface, but the size of the borings makes the subsistence of the organisms forming them seem improbable on a diet of soil microfauna and microalgae. However, the possibility exists that the burrows or borings in this Ordovician section pre-date the weathering of the soil profile which identifies its subaerial origin. Alternatively, they might represent the work of aquatic organisms which burrowed during occasional incursions of water over the soil surface, in the overbank deposit in which they occur (Wright 1990).

3. DEVONIAN EVIDENCE OF PLANT–ARTHROPOD INTERACTIONS

The Rhynie Chert of early Devonian (Siegenian–Emsian) age from Aberdeenshire represents a peat-bog ecosystem that was repeatedly inundated by hot volcanic fluids that fixed the plant and animal tissue, and subsequently infiltrated it with near-transparent silica (Kidston & Lang 1921). This remarkable preservation of a wetland ecosystem gives us a picture of the participants in a Devonian food web involving algae, tracheophytes and several other enigmatic, probably land-adapted, plants together with a number of different types of arthropod, some aquatic, some evidently terrestrial. The arthropods include an aquatic crustacean, the earliest record of a hexapod in the form of a collembolan, *Rhyniella praecursor* Hirst and Maulik, mites, including *Protacarus crani* Hirst and the largest of all the Rhynie arthropods, the trigonotarbid arachnid *Palaeocharinus scourfieldi* Hirst. This has booklungs, as a clear adaptation to terrestrial existence (Jeram *et al.* 1990). It is a reasonable first working hypothesis, based on their nearest living counterparts, that the mites and collembolan were litter feeders, and that the arachnid was zoophagous, and preyed on the smaller arthropods (and presumably any other soft-bodied fauna available to it, of which no fossil remains may have survived). This very conjectural sketch of a food web carries the implication of the underlying terrestrial primary productivity by the plants, upon which this community must have been dependent.

There are two pieces of more tangible evidence which point to plant–animal relations in this early terrestrial ecosystem. Some of the axes in the Rhynie Chert show various types of lesions. These were noted by the first authors to describe the Rhynie plants, Kidston & Lang (1921), and they attributed them to physical injury caused by the circumstances of the (volcanic-based) fossilization process. However, some features of these lesions make this seem improbable. Notch-like lesions extend inwards to the vascular tissue, and specifically to the phloem-like tissue surrounding the xylem. The lesions are filled with a dark plug of ?exudate. Most significantly, the cells abutting on the break in the tissue have expanded into the gap, and have undergone division after the injury occurred. The exudation and the wound response of cell division are typical reactions of a living plant to injury. Patently the lesions were made while the plant was alive, and the plant survived the injury (Kevan *et al.* 1975). It is at least possible that the injury was caused deliberately to gain access to sap exudation as a means of feeding. However, Rolfe (1985 *a*) queries the biological cause for these lesions. Of the known arthropods in the fauna, Kevan *et al.* (1975) suggest *Protacarus* and *Rhyniella* as possible causal organisms. In view of the size of the lesions, it must be acknowledged that the much larger trigonoitarbids would have been better equipped to make such injuries. Other biological causal agents – nematodes for example, which might leave no other tangible trace of their presence – are certainly possible, and cannot be discounted (Conway-Morris 1981; Wallace 1973). The possibility of a sap-feeding relation involving the arthropods is probably about as far as this evidence can take us at present.

The suggestion of a different relation in the Rhynie biota exists in the occurrence of trigonotarbids inside empty *Rhynia* sporangia (Rolfe 1980, 1985 *a*; Kevan *et al.* 1975). The sporangial walls of *Rhynia* are of a distinctive 'prismatic' character, so that it is very clear in Rolfe's Rhynie material that the trigonotarbids were within a sporangial cavity. Their presence in that particular location may of course have been quite

Table 1. *Results of simple feeding experiments on the viability of bracken spores after being eaten by three types of terrestrial arthropod*

(The first line of figures are from the results given by Chaloner (1976); a similar feeding protocol was used for the results given here, but the grass used as feed in the earlier experiments was here replaced by bran. For each of the three arthropod types, their bran was dusted with spores of *Pteridium aquilinum*, the arthropods then separated from their food and the droppings collected for 24 h. The droppings were then dispersed in a small volume of water, plated out on mineral agar and, after a week, all spores observed were scored as having germinated or not, and this expressed as a percent viability (last column). A control germination rate was also scored on spores plated out directly (first column).

The much lower germination rate seen in the current *Locusta* feeding is presumably due at least in part to the different degree of mastication involved in eating spores associated with bran rather than the grass used in the earlier experiments.)

arthropod	% viable in control	ungerminated	germinated	% viable in droppings
Locusta migratoria (Chaloner 1976)	67	—	—	47
Locusta migratoria	74	101	7	6
Spirobolus sp. (giant African millipede)	74	63	5	7
Periplaneta americana (cockroach)	74	100	0	0

fortuitous, as he concedes. However, they might have entered the empty sporangia as a means of obtaining shade or shelter from desiccation. A further possibility is that they might have eaten the contents of the sporangium before occupying it. The biological significance of such spore eating is considered further below in a Carboniferous context, where the evidence for arthropod spore eating is very much stronger.

In addition to the evidence from the Rhynie occurrence, Banks (1981) has published illustrations of the leafless early Devonian plant *Psilophyton* from the Lower Devonian of Gaspe showing injuries consistent with stem feeding, either by arthropods or some other organism. Here, the wound reaction is more clearly defined than in the *Rhynia* axes, with a periderm-like response of cell division over quite a wide zone beneath a region of superficial damage. As with the Rhynie material we can only speculate on a biological causal agent, largely because it is so hard to picture any purely physical cause for the injury that Banks reports.

4. CARBONIFEROUS EVIDENCE FOR PLANT–ARTHROPOD INTERACTIONS

The exploitation of Northern Hemisphere Carboniferous coals has brought a great deal of fossil-bearing rock of that age to the attention of scientists over the last century and a half. For this reason alone it is not surprising that our record of plant–arthropod interactions within the Carboniferous is more diverse and better documented than from the Devonian. But even allowing for this, it is clear that an expansion of terrestrial life within the Carboniferous gave scope for a wider range of interactions. For within this period we have record of arthropod gut contents including plant material, of fossil droppings (coprolites) containing abundant spore exines, together with evidence of wood boring and, by the late Carboniferous, some evidence of leaf eating. There is also sound documentation of a steady increase in the size of plant propagules (megaspores and seeds) which represent an exploitable

plant product of high nutritive value, coupled with evidence of a wider range of land-adapted arthropods including winged insects. Carboniferous evidence for plant animal interactions have been extensively reviewed (Hughes & Smart 1967; Rolfe 1985 *b*; Scott & Taylor 1983), and we comment on a limited number of instances to illustrate the range of evidence available.

The most convincing record of plant eating by arthropods must of course be the presence of plant remains in their gut contents. Rolfe & Ingham (1967) have illustrated fragmentary xylem, probably of a lycopod, in the gut of the large arthropod, *Arthropleura*, from the late Carboniferous. Scott has prepared fossil spore exines, probably also of lycopod origin, from the gut contents of the winged Carboniferous insect *Eucaenus* from the famous Mazon Creek locality in Illinois (Scott & Taylor 1983). These occurrences are important as they are the most convincing evidence possible for phytophagy, and yet it must also be borne in mind that spore exines and lignified cells of xylem represent two of the most indestructible forms of plant tissue. Their survival through the process of digestion, and subsequent fossilization, would favour their occurrence in any remaining gut contents, even if their ingestion had been quite fortuitous, or formed only a minor dietary component. Even a carnivorous animal living in contemporary grassland in early summer, as a herbivore predator, might well consume quite a lot of grass pollen, either first or second-hand. The presence of a small quantity of grass pollen in the gut of any animal would be a frail basis for deducing that pollen eating was a significant part of the dietary pattern!

This gentle caution can surely be set aside in the case of the late Carboniferous droppings described by Scott (1977) consisting largely of fossil spore exines (figures 8 and 9). The attribution of any such coprolites to any given source animal is inevitably problematical, but as Scott has argued, there is a strong probability that these droppings are of terrestrial arthropod origin. Although the spore content is diverse, it is so pervasive in these droppings that the suggestion that the

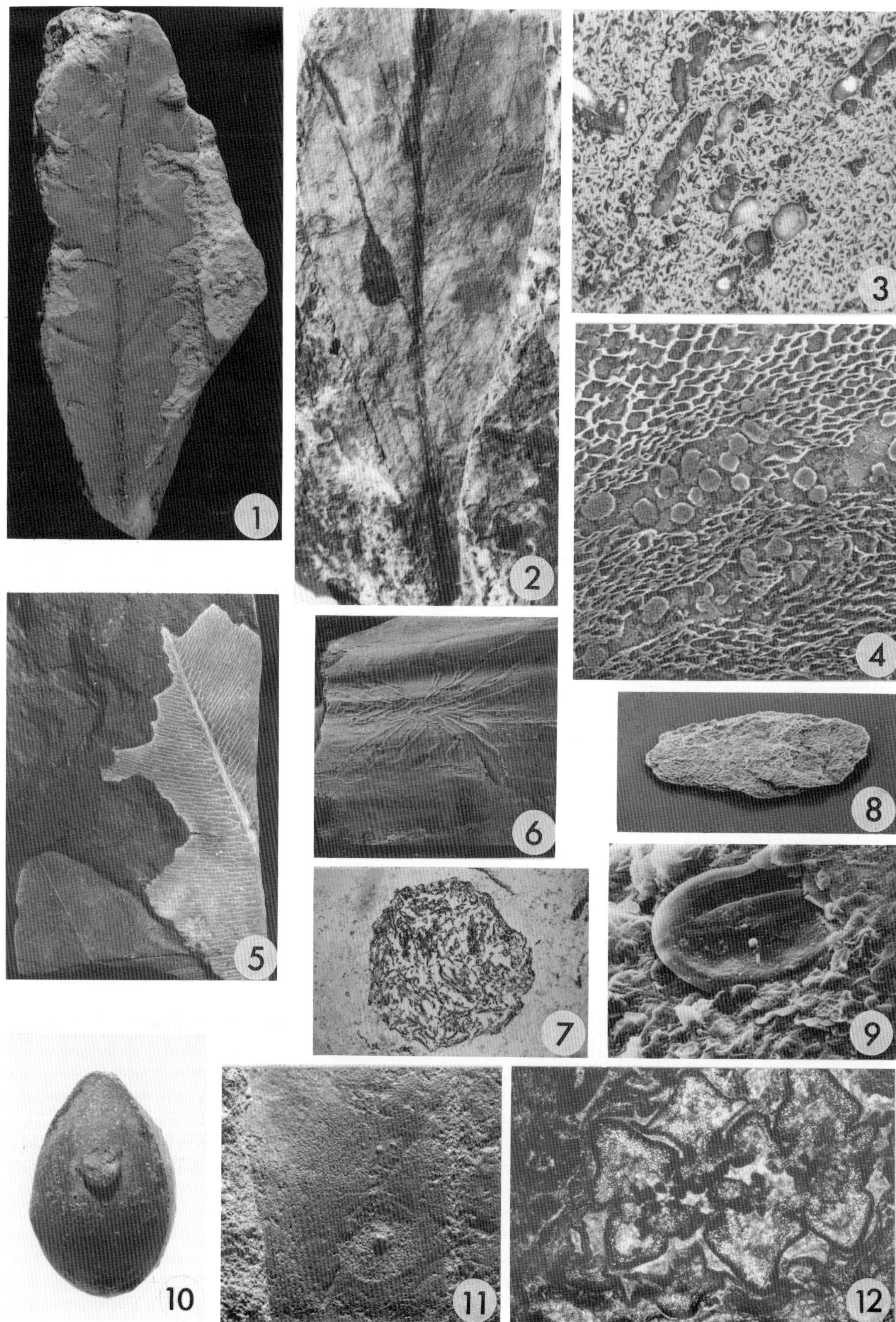

Figures 1–12. For description see opposite.

organism producing them was a spore eater seems inescapable. None the less, as noted above, it is also possible that an arthropod was feeding on some other organism which was itself spore eating, or that it was taking in a high proportion of spores through feeding on spore-rich litter.

With this proviso about the start of the spore-eating food-chain, it is worth noting that the nutritional value (and especially the protein content) of pollen is comparable to that of seeds, (which are generally speaking the most rewarding of plant tissues in the modern flora), and spores of pteridophytic plants are probably of similar composition (Scott *et al.* 1985). It is significant that Southwood (1973), on the basis of comparative studies of living insects (and without reference to the fossil record) wrote that 'the evolutionary path of the phytophagous insect has not been an easy one; pollen feeding often seems to represent "the first step" and feeding on foliage "full success". One might add that spore eating was the first step, which led not only to pollen eating as the seed plants evolved, but that this particular plant–animal interaction was in turn the pathway into biotic pollination in the flowering plants.

The probable significance of spore eating as a lead into insect pollination makes it appropriate to reflect on how far this relationship might be as mutually advantageous to the spore-bearing plant and its spore predator, as is biotic pollination to both flowering plant and vector. As Chaloner (1976) and Scott *et al.* (1985) have suggested, spores able to survive passage of the gut would then be deposited in the moist and sheltered environment of the insect's faecal pellet, after

having been transported away from the parent plant (perhaps some considerable distance, in the case of a flying insect). Unlike the randomly deposited wind-dispersed spore, a spore transported in this way would presumably end up in an environment at least acceptable to the insect. In that the insect 'predator' would thus have assisted in the survival and spread of its own food source plant, the relationship would obviously be mutually advantageous.

Chaloner (1976) did some simple experiments on the survival rate of *Pteridium* spores fed to locusts. (The insect order Orthoptera, to which the locusts belong, is known from the Carboniferous onwards.) He showed that passage through the locust gut only reduced spore viability by about 50%. We have repeated these experiments, using millipedes and cockroaches, in addition to locusts (see table 1). The choice of these arthropods was influenced by the occurrence of the Class Diplopoda, and the insect Order Blattodea (to which these two groups respectively belong) also being amongst the fossil forms known from the Carboniferous (Scott *et al.* in preparation). Although the cockroach spore feeding gave no viable spores in the droppings, the spore survival rate in the millipede gut was actually higher than that in the locust. The feeding protocol and the procedure for collecting the droppings, uncontaminated by the spore-dusted food, were essentially the same as those that Chaloner had used, and the fern spores had a higher initial viability than in his material. Even so, the new locust results gave a much lower survival rate than in the earlier experiments. It is not clear just why these later feeding experiments produced a higher mortality in the fern spores eaten by

---

Figure 1. Partially eaten angiosperm leaf showing marginal and rare non-marginal feeding (notably, *ca.* 2 cm up leaf near left-hand margin) from the Ripley Formation (Maastrichtian, Cretaceous), U.S.A. (PP11525, Field Museum of Natural History, Chicago, U.S.A.) ($\times 1\frac{1}{2}$.)

Figure 2. Trumpet type leaf mine in angiosperm leaf. The mine originates near the margin then proceeds along a secondary vein until it forms a relatively large blotch-like pupation chamber which is full of frass. There is a central exit pore. Dakota Formation (Cenomanian, Cretaceous). (Indiana University Collection, IU15703-2529, U.S.A. Courtesy of D. Dilcher.) ($\times 2$.)

Figure 3. Possible arthropod borings in gymnospermous wood preserved as fusain, late Lower Carboniferous, East Kirkton, Scotland. (Scott Collection.) ($\times 100$.)

Figure 4. Gymnospermous wood with borings filled with frass from the Upper Carboniferous of the U.S.A. (Botany Department, Ohio State University, Columbus Ohio, U.S.A.) (Scanning electron micrograph $\times 60$.)

Figure 5. Partially eaten leaf of *Glossopteris* from the Permian of Australia, showing continuous marginal feeding traces. (Scott Collection.) ($\times 1$.)

Figure 6. Traces of a bark-burrowing beetle on a gymnosperm log from the Lower Cretaceous of southern England. (v7535, Natural History Museum, London, U.K.) ($\times \frac{1}{2}$.)

Figure 7. Cross section of coprolite containing plant material from the Kingswood Limestone, Lower Carboniferous of Pettycur, Fife, Scotland. (Scott Collection.) ($\times 10$.)

Figure 8. Coprolite containing plant material, from the Upper Carboniferous of Yorkshire. (FSC2063, Hunterian Museum, Glasgow, U.K.) (Scanning electron micrograph $\times 15$.)

Figure 9. Detail of Upper Carboniferous coprolite from Yorkshire showing contents of spores and pollen. (FSC 2068, Hunterian Museum, Glasgow, U.K.) (Scanning electron micrograph $\times 250$.)

Figure 10. Sandstone cast of seed (*Trigonocarpus*) with plug representing hole in original seed coat. Upper Carboniferous. (Geology collections, unregistered, National Museum of Scotland, U.K.) ($\times 1\frac{1}{2}$.)

Figure 11. Angiosperm leaf with spot gall with central exit hole, from the Dakota Formation (Cenomanian, Cretaceous), U.S.A. (UP348, Field Museum of Natural History, Chicago, U.S.A.) ($\times 2$.)

Figure 12. Small coprolites (possibly of mite) in lycopod axis, from the Lower Carboniferous Pettycur Limestone, Pettycur, Fife, Scotland. (Gordon Collection No. 176, Natural History Museum, London, U.K.) ($\times 15$.)

locusts, but the point of real significance is that spores, in what ever proportion, can survive passage through the gut of two such voracious plant eaters as the locust and the millipede.

If, for every hundred spores eaten, five are deposited well away from the parent source, in a moist humus-rich environment, presumably in a habitat suitable for both plant and spore-eater, then the exercise may be seen as well worth the cost of the 95 spores lost to the plant. In angiosperm biotic pollination, a pollen production of a thousand grains per ovule produced must be rated as being at the efficient end of the range (Faegri & Van der Pijl 1971). It seems that spore eating, coupled with ensuing dispersal (which might yield successful gametophytes, rather than the viable seed of a successful pollination) would be a worthwhile process to the spore-bearing plant. These simple experiments at least give some support to the suggestion that spore eating would have been of mutual benefit to both plant and spore eater. This is of course simply a special case of the process long-recognized as endo-zoochory, in which seed dispersal is effected by many fruit-eating animals. In that case it is the seeds, which are often specially adapted to survive passage through the processes of eating and digestion, which are dispersed when they are voided, still in a viable state, by the animal.

Coprolites formed of plant debris, other than those containing spores, have been described from the early Carboniferous of Scotland and France (Scott 1977; Scott *et al.* 1985; Rex & Galtier 1986; figures 7 and 12). The latter authors recognize four different types of coprolites, which they regard as evidence for some diversity in the phytophagous fauna, some of these occurring within plant tissue, and others dispersed within the matrix.

There is other evidence from within the Carboniferous of arthropods eating plant propagules (spores, seeds) without any element of a mutual benefit. The megaspores of some of the tree-sized lycopods of the Carboniferous evidently represented a nutritional source worth tapping, and megaspores have been reported showing a neat borehole through which feeding presumably had occurred, much as in modern insect damage to seeds (Scott & Taylor 1983). Seeds of Carboniferous age showing (presumed) arthropod borings have been reported by several authors. We illustrate an internal cast (i.e. a sediment infill of the stony integument) of the Carboniferous seed *Trigono-carpus* (figure 10). The boring appears as a sediment plug, protruding from the cast, as matrix filled both seed cavity and the bored hole through the integument. It is not clear from the nature of the hole in either megaspore wall or seed integument whether it was made to facilitate access for feeding, or whether it represents emergence of a mature or maturing insect or other arthropod, after egg-laying within the plant propagule. In the two instances cited, the former case seems more likely for the megaspore, on the basis of the relatively large size of the hole, whereas the seed boring could be plausibly explained on either basis.

The evidence from the Carboniferous spore-rich coprolites certainly pre-dates any abundant evidence of leaf eating, although leaves with bite holes are known from the late Carboniferous. However, the number of records is absurdly low, in relation to the number of fossil leaves of Carboniferous age in museums, and reported in the literature. Scott & Taylor (1983) in a survey of a well-worked late Carboniferous flora from North America reported only 4% of one of the commonest leaf species present showing evidence of leaf feeding. It must be accepted that the apparent rarity of Carboniferous leaves showing feeding damage may be a result of such 'imperfect' fossil specimens being deliberately – or perhaps even subconsciously – ignored by collectors. But even acknowledging this possible collecting bias, Southwood's suggestion quoted above is certainly consistent with the evidence of spore eating preceding any extensive evidence of leaf feeding within the Carboniferous.

It has become the practice of palaeobiologists working on a wide range of 'trace fossils' (ichnofossils, such as partially eaten leaves or footprints, in which the causal animal leaves no actual fossil remains of itself), to apply formal binomial names to the traces. These so-called ichnotaxa are nomenclaturally somewhat curious. We have, for example, the genus *Phagophytichnus* (see, for example, Van Amerom (1966)), which is applied to the trace fossil comprising bite marks along the edges of leaves, from Palaeozoic to Tertiary in age (see figures 1 and 5). The name relates of course to the somewhat elusive concept of the unknown animal that did the biting. The nomenclatural type-specimen of such a 'trace fossil' is the plant fossil showing the bites, or perhaps more exactly, the missing parts of the plant fossil, resulting from the biting. None the less, these ichnofossil names form a useful language to record the evidence of the important process of leaf feeding.

We have well-documented evidence of marginal leaf feeding from the late Carboniferous with *Phago-phytichnus* bites on the leaves of the seed-fern *Neuropteris* (Van Amerom 1966; Scott & Taylor 1983). Such evidence of marginal leaf feeding has been recorded more frequently in the Permian (figure 5), and is considered below.

Our earliest record of wood boring also comes from within the Carboniferous (fig. 3), even though sizeable stems with extensive secondary xylem occur from the mid-Devonian onwards. The wood-boring habit has, of course, the dual role for the borer in combining 'feeding' with 'protection' from predators, within the bulk of the plant tissue. However, unless it has some form of cellulose-digesting microflora in its gut, the bulk of wood consumed by a boring insect is unusable nutritionally. The mites and collembolans of the early Devonian were presumably detritivores, consuming plant material already partially degraded by microbial activity (fungal, bacterial) in the soil litter. We have at least some well-documented evidence of fungal activity within the tissue of the Rhynie plants (Kidston & Lang 1921) even though later authors have favoured a symbiotic basis for their role rather than a saprophytic one. It seems an obvious progression from this re-lationship of arthropod–soil-microflora–litter-feeding, to one of arthropod–gut-microflora–plant-feeding.

This would represent the evolution from a casual relationship of arthropod and cellulose-degrading microorganisms to one of mutual advantage between the arthropod and its gut microflora.

The first evidence of any considerable wood-boring activity is none the less in late Carboniferous gymnosperm wood, with 'galleries' containing coprolites of the boring organisms described by several authors (figure 4). Orabatid mites are favoured as likely borers (Cichan & Taylor 1982; Scott & Taylor 1983). Several examples of insect boring within fern tissue have also been reported (Rothwell & Scott 1983; Lesnikowska 1990). The latter is important in showing a wound response associated with the boring, indicating that the plant was alive when the feeding took place. There are a number of records of wood boring (and even bark burrowing, figure 6) in Mesozoic gymnosperm material, with a wider range of possible causal organisms, and these are reviewed in Scott *et al.* (in preparation).

Leaf mining is a further type of insect attack on plants which combines 'feeding' with 'protection' (fig. 2). The majority of modern leaf miners are found in angiosperm leaves, where the causal insect lays its eggs upon or within the leaf and the larva then feeds on the mesophyll, leaving a trail within the leaf that is recognizable in the fossil state. The earliest record of leaf mining is in late Carboniferous specimens of the seed-fern genus *Neuropteris* (Muller 1982), which show blotches and linear markings. These and somewhat younger leaf mines described by the same author are said to show similarity to those produced by modern lepidopteran miners, but fossil insect remains of that group do not appear until the Cretaceous.

5. POST-CARBONIFEROUS PLANT–INSECT INTERACTIONS

All the types of tangible evidence of plant–arthropod interactions reported above are known in post-Carboniferous material. In the Permian we have further evidence of leaf eating with *Phagophytichnus*-type marginal feeding from the Southern Hemisphere gymnosperm *Glossopteris* (Plumstead 1963; Stephenson & Scott 1991) (figure 5). Mesozoic records of leaf eating are fairly sparse, but they increase in number and diversity of feeding strategies as the angiosperms come to dominate the terrestrial flora towards the close of the Cretaceous. In the early angiosperms we see a more extensive record of 'continuous marginal' and 'interrupted marginal' leaf feeding, and eventually the appearance of 'non-marginal' feeding: the taking of isolated patches of tissue within the body of a leaf (Stephenson, unpublished data) (figure 1). The latter is evidently a more difficult feeding process for insects, and significantly, we have no record of it earlier than the Cretaceous.

Inevitably the fossil record gives little indication of what was probably the most prevalent interaction between plants and animals, namely the 'chemical warfare' involved in the plant kingdom seeking respite from the plant eater. This, more than any other interaction short of biotic pollination, must have been in the nature of a coevolutionary response as the production of antifeedant or toxic defence substances by the plant was countered by some physiological modification of the plant-feeding animal. The seemingly trivial differences of leaf feeding strategy probably reflect aspects of this complexity of the plant–animal and herbivore–predator biochemical relation, in ways only hinted at in the simple observation of damaged fossil leaves. Edwards & Wratten (1980, 1983) give a helpful review of the significance of various feeding patterns. The production of antifeedant substances in immediate response to leaf damage is doubtless connected with discontinuity of marginal feeding, and probably also to 'internal' feeding. We illustrate an example of a Cretaceous leaf showing one of the earliest records of such 'internal' feeding damage in an angiosperm (figure 1). Avoiding the advertisement to predators offered by extensive and conspicuous leaf feeding is clearly also a factor governing feeding behaviour, but of this we can have no direct evidence from fossil leaves showing feeding damage.

One of the most elaborate and biochemically finely-tuned interplays of the plant–animal relation is that of plant galls induced by a parasite. This response of plant tissue, in abnormal growth caused by substances produced by the invading parasite, may give the animal both protection against predators, and an enhanced food supply for a developing larva. Our earliest record of plant gall formation is in the Permian report of Potonié (1893) of leaf galls on the pteridosperm *Odontopteris* (see also Conway-Morris (1981)). This isolated Palaeozoic occurrence is followed by a scattered record of Mesozoic galls (Scott *et al.*, in preparation). With the rise of the angiosperms from the mid-Cretaceous, a wide range of leaf galls gives increasing evidence of this type of insect activity. Stephenson has explored mid-Cretaceous angiosperm leaf floras for evidence of gall formation, and recognizes 26 leaf specimens with galls, which fall into eight distinct categories. An example of one of these is illustrated in figure 11. The range of forms is consistent with those formed by present-day Acari, Hemiptera, Diptera and Hymenoptera. It is a little surprising that none are clearly attributable to the Lepidoptera, as this group has a fossil record from the Cretaceous, and is likely to have been involved in insect–angiosperm coevolution in the role of vectors of biotic pollination.

Leaf mining has a record that runs from the Carboniferous, as reported above, through records in Permian and Mesozoic gymnosperm leaves (see Scott *et al.*, in preparation). Angiosperm diversification – and perhaps, in particular, the soft leaf texture of their deciduous forms – seems to have given added scope to this strategy. One of us (J.S.), in a study of Cretaceous angiosperm leaves has found 50 occurrences, representing 18 different types of leaf mine patterns. In some cases the mines are so well preserved that they trace the growth of the expanding larva from the site of entry to the pupation chamber, where the insect finally vacates the leaf (figure 2; see Hering 1951).

Spore feeding also has a Mesozoic record in the form of spore-containing coprolites. Harris (1957) gives one of the earliest reports of small coprolites containing the pollen of the gymnosperm *Caytonia* from the Middle Jurassic. In terms of plant–animal coevolution, this direct evidence of some small animal engaged in pollen feeding is inevitably to be seen in the context of the biotic pollination that characterizes some of the primitive (and earliest) angiosperms. The role of pollen feeding as the lead into repeated flower-visiting, and hence into a mutually advantageous flower–vector relationship has been amply expounded on the basis of observation of living plants (Faegri & Van der Pijl 1971), and is discussed elsewhere in this symposium (Friis & Crepet).

We thank Deborah Parsons and Geoffrey Prior for their help with the modern arthropod experiments, and Kevin de Souza for photographic help. J. S. gratefully acknowledges the receipt of a NERC research studentship.

REFERENCES

Banks, H. P. 1981 Peridermal activity (wound repair) in an Early Devonian (Emsian) trimerophyte from Gaspé Peninsula, Canada. *Palaeobotanist* 28–29, 20–25.

Chaloner, W. G. 1976 The evolution of adaptive features in fossil exines. In *Evolutionary significance of the exine* (ed. I. K. Ferguson & J. Muller), pp. 1–14. London: Academic Press.

Chaloner, W. G. 1988 Early land plants – the saga of a great conquest. In *Proceedings of the XIV International Botanical Congress* (ed. W. Greuter & B. Zimmer), pp. 301–316. Konigstein/Taunus: Koeltz.

Cichan, M. A. & Taylor, T. N. 1982 Wood-borings in *Premnoxylon*: plant-animal interactions in the Carboniferous. *Palaeontol. Palaeoclimatol. Palaeoecol.* 39, 123–127.

Conway-Morris, S. 1981 Parasites and the fossil record. *Parasitology* 82, 489–509.

Edwards, D. & Burgess, N. D. 1990 Terrestrialization: Plants. In *Palaeobiology: a synthesis* (ed. D. E. G. Briggs & P. R. Crowther), pp. 60–64. Oxford: Blackwells.

Edwards, D., Fanning, U. & Richardson, J. D. 1986 Stomata and sterome in early land plants. *Nature, Lond.* 323, 438–440.

Edwards, D. & Feehan, J. 1980 Records of *Cooksonia*-type sporangia from late Wenlock strata in Ireland. *Nature, Lond.* 287, 41–42.

Edwards, P. J. & Wratten, S. D. 1980 Ecology of insect–plant interactions. *Inst. Biol. stud. Biol.* 121.

Edwards, P. J. & Wratten, S. D. 1983 Wound induced defences in plants and their consequences for patterns of insect grazing. *Oecologia* 59, 88–93.

Faegri, K. & Van der Pijl, L. 1971 *The principles of pollination biology*, 2nd edn. (291 pages.) Oxford:

Gray, J. 1985 The microfossil record of early land plants: advances in understanding of early terrestrialization. In *Evolution and environment in the late Silurian and early Devonian* (ed. W. G. Chaloner & J. O. Lawson) (*Phil. Trans. R. Soc. Lond.* B 309), pp. 167–195.

Harris, T. M. 1957 How we study fossil plants: *Caytonia*. *New Biol.* 22, 24–38.

Hering, E. M. 1951 Biology of leaf miners. (420 pages.) s'Gravenhage, Netherlands: Junk.

Hueber, F. M. 1983 A new species of *Baragwanathia* from the Sextant Formation (Emsian), Northern Ontario, Canada. *Bot. J. Linn. Soc.* 86, 57–79.

Hughes, N. F. & Smart, J. 1967 Plant-insect relationships in Palaeozoic and later time. In *The fossil record* (ed. W. B. Harland), pp. 107–117. Geology Society of London.

Jeram, A. J., Selden, P. A. & Edwards, D. 1990 Land animals in the Silurian: arachnids and myriapods from Shropshire, England. *Science, Wash.* 250, 658–661.

Kevan, P. G., Chaloner, W. G. & Savile, D. B. O. 1975 Interrelationships of early terrestrial arthropods and plants. *Palaeontology* 18, 391–417.

Kidston, R. & Lang, W. H. 1921 On Old Red Sandstone plants showing structure, from the Rhynie Chert Bed, Aberdeenshire. Part V. The thallophyta occurring in the peat bed; the succession of the plants through the vertical section of the beds, and the conditions of accumulation and preservation of the deposit. *Trans. R. Soc. Edinb.* 51, 855–902.

Lesnikowska, A. D. 1990 Evidence of herbivory in tree-fern petioles from the Calhoun Coal (Upper Pennsylvanian) of Illinois. *Palaios* 5, 76–80.

Muller, A. H. 1982 Über Hyponomefossiler und rezenter Insekten, erster Beitrag. *Freiberger Forschungsheft* C 366, 7–27.

Plumstead, E. P. 1963 The influence of plants and environment on the developing animal life in Karoo times. *S. Afr. J. Sci.* 59, 147–152.

Potonié, H. 1893 Die Flora des Rothliegenden von Thuringen. Abhandlungen der Koniglich Preussischen geologischen Landesanstalt 9, 1–298.

Retallack, G. & Feakes, C. 1987 Trace fossil evidence for Late Ordovician animals on land. *Science, Wash.* 325, 61–63.

Rex, G. M. & Galtier, J. 1986 Sur l'évidence d'interactions animal-vegetal dans le Carbonifère inférieur Français. *C. r. Acad. Sci., Paris* Ser 303 (II), 17, 1623–1626.

Robbins, E. I., Porter, K. G. & Haberyan, K. A. 1985 Pellet microfossils: possible evidence for metazoan life in Early Proterozoic time. *Proc. natn. Acad. Sci. U.S.A.* 82, 5809–5813.

Rolfe, W. D. I. 1980 Early invertebrate terrestrial faunas. In *The terrestrial environment and the origin of land vertebrates* (ed. A. L. Panchen), pp. 117–157. London: Academic Press.

Rolfe, W. D. I. 1985*a* Early terrestrial arthropods: a fragmentary record. *Phil. Trans. R. Soc. Lond.* B 309, 207–218.

Rolfe, W. D. I. 1985*b* Aspects of the Carboniferous terrestrial arthropod community. *Proceedings of the IXth International Congress of Carboniferous Stratigraphy and Geology (Urbana, 1979)*, 5, 303–316.

Rolfe, W. D. I. 1990 Seeking the arthropods of Eden. *Nature, Lond.* 348, 112–113.

Rolfe, W. D. I., Durant, G., Fallick, A. E., Hall, A. J., Large, Scott, A. C., Smithson, T. R. & Walkden, G. 1990 An early terrestrial biota preserved by Visean vulcanicity in Scotland. In *Volcanism and fossil biotas* (ed. M. Lockley & A. Rice) *Geol. Soc. Am., spec. Pub.* 244, 13–24.

Rolfe, W. D. I. & Ingham, J. K. 1967 Limb structure, affinity and diet of the Carboniferous "centipede" Arthropleura. *Scott. J. Geol.* 3, 118–124.

Rothwell, G. W. & Scott, A. C. 1983 Coprolites within the marattiaceous fern stems (*Psaronius magnificus*) from the Upper Pennsylvanian of the Appalachian Basin, U.S.A. *Palaeogeog. Palaeoclimatol. Palaeoecol.* 41, 227–232.

Scott, A. C. 1977 Coprolites containing plant material from the Carboniferous of Britain. *Palaeontology* 20, 59–68.

Scott, A. C., Chaloner, W. G. & Paterson, S. 1985 Evidence of pteridophyte–arthropod interactions in the fossil record. *Proc. R. Soc. Edinb.* B 86, 133–140.

Scott, A. C. & Paterson, S. 1984 Techniques for the study of plant/arthropod interactions in the fossil record. *Geobios Mem. Spec.* 8, 449–455.

Scott, A. C., Stephenson, J. & Chaloner, W. G. Fossil evidence for plant–animal (arthropod) interaction and coevolution in the Palaeozoic and Mesozoic. (In preparation.)

Scott, A. C. & Taylor, T. N. 1983 Plant/animal interactions during the Upper Carboniferous. *Bot. Rev.* **49**, 259–307.

Sherwood-Pike, M. A. & Gray, J. 1985 Silurian fungal remains: probable records of the class Ascomycetes. *Lethaia* **18**, 1–220.

Southwood, T. R. E. 1973 The insect/plant relationship – an evolutionary perspective. In *Insect/plant relationships* (ed. H. F. Van Emden) (*Symp. R. ent. Soc. Lond.* **6**), pp. 3–30.

Stephenson, J. & Scott, A. C. The geological history of insect related phytopathology. (In preparation.)

Trant, C. A. & Gensel, P. G. 1985 Branching in *Psilophyton*: a new species from the Lower Devonian of New Brunswick, Canada. *Am. J. Bot.* **72**, 1256–1273.

Van Amerom, H. W. J. 1966 *Phagophytichnus ekowsskii* nov. ichnogen. & nov. ichnosp., eine missbildung infolge von insektenfrass, aus dem Spanischen Stephanien (Provinz Leon). *Leidse Geologische Mededelingen* **38**, 181–184.

Wallace, H. R. 1973 *Nematode ecology and plant disease.* London: Edward Arnold.

Wright, V. P. 1990 Terrestrialization: soils. In *Palaeobiology – a synthesis* (ed. D. E. G. Briggs & P. R. Crowther), pp. 57–59. Oxford: Blackwells.

Discussion

E. A. Jarzembowski (*Booth Museum of Natural History, Brighton, U.K.*). Please could the authors consider providing criteria to distinguish insect or other arthropod feeding marks on fossil leaves as Crane & Jarzembowski (1990) have attempted for leaf mines. This is to distinguish them from the vagaries of geological preservation and damage from other physical or biotic causes. Necrotic tissue provides evidence of damage during the life of a plant but, like fungal attack, may help obscure important evidence, for example the jagged edge left on a leaf by a mandibulate insect.

Reference

Crane, P. R. & Jarzembowski, E. A. 1990 Insect leaf mines from the Palaeocene of southern England. *J. Nat. Hist.* **14**, 629–636.

W. G. Chaloner. The best indication we can have that damage to a fossil leaf has occurred while it was alive is the presence of necrotic tissue at the margin of the injury. We discuss this matter in further detail in Scott *et al.* (1991).

E. A. Jarzembowski. *Rhyniella praecursor* (Collembola: Isotomidae) from the Rhynie Chert (Lower Devonian: Pragian), the earliest known insect (hexapod), is now known from the whole anatomy, the previously unknown hindbody having been described by Whalley & Jarzembowski (1981). A new reconstruction of *R. praecursor* is modelled by Jarzembowski (1989, figure 5).

References

Jarzembowski, E. A. 1989 A century plus of fossil insects. *Proc. Geol. Ass.* **100** (4), 433–449.

Whalley, P. E. S. & Jarzembowski, E. A. 1981 A new assessment of *Rhyniella*, the earliest known insect, from the Devonian of Rhynie, Scotland. *Nature, Lond.* **291**, 317.

W. G. Chaloner. We refer to the 1981 record in this paper, and are pleased to learn of the new reconstruction of *Rhyniella*.

E. A. Jarzembowski. An important element of the Upper Carboniferous fauna was the presence of rostrate palaeopterous insects that may have occupied the ecological niche later held by hemipteroid insects (Jarzembowski 1987). Could the pierced *Trigonocarpus* and spores be the feeding traces of these insects as suggested by A. G. Sharov and P. Barnard (Wootton 1981)?

References

Jarzembowski, E. A. 1987 The occurrence and diversity of Coal Measure insects. *J. geol. Soc. Lond.* **144**, 507–511.

Wootton, R. J. 1981 Palaeozoic insects. *A. Rev. Ent.* **26**, 319–344.

W. G. Chaloner. Yes, this is possible; see also Shear & Kukalova-Peck (1990).

Reference

Shear, W. A. & Kukalova-Peck, J. 1990 The ecology of Paleozoic terrestrial arthropods: the fossil evidence. *Can. J. Zool.* **68**, 1807–1834.

E. A. Jarzembowski. The beetle-bored wood from the English Wealden is of additional interest because it shows two types of borings (Jarzembowski 1990) supporting the view that a stressed tree is liable to attack by more than one vector. Bark beetles, which have been considered responsible for some of these borings belong to the same family as weevils (Crowson 1981) which are now being suggested; could the burrows have been produced by larvae belonging to stem-group Curculionidae?

References

Crowson, R. A. 1981 *The biology of the Coleoptera* (802 pages.) London: Academic Press.

Jarzembowski, E. A. 1990 A boring beetle from the Wealden of the Weald. In *Evolutionary paleobiology of behaviour and coevolution*, (ed. A. Boucot), pp. 373–376. Elsevier.

W. G. Chaloner. Yes, of course, this too is possible.

R. McN. Alexander (*Department of Pure and Applied Biology, Univrsity of Leeds, U.K.*). I am interested by the suggestion that spores may be a rewarding food material although a substantial proportion pass through the gut unharmed. Does this imply that digestion of a spore is very much an all-or-nothing matter: that a spore either breaks open and is thoroughly digested, or survives intact and unharmed?

W. G. Chaloner. Yes, it does seem to be a more or less hit-or-miss affair: that is, any particular spore is either ruptured, presumably in mastication, or it is not. In the latter case, it has a chance of surviving and germinating. As a result we see three categories of spores in the droppings: (i) those which germinate (and which we score as such); (ii) those which are ruptured and empty and (iii) those which appear intact, but which have not germinated. This last category may have already been in a non-viable state when they were eaten, or may have been killed during or after eating, but without being ruptured. In any event, we scored these two latter categories together as being non-viable in the droppings. We could (and perhaps should) have disoriminated between these, and scored them separately.

It is a little surprising that quite a significant fraction do pass the gut unharmed. But it seems likely that the proportion is influenced by the nature of the food on which they are fed to the arthropod, and hence presumably to the degree of mastication to which they are subjected. The entire process is obviously susceptible to much more serious experimentation than we have attempted!

M. E. COLLINSON (*Biosphere Sciences Division, Kings College London, U.K.*). Do the authors have evidence that herbivores change their diet to exploit new groups of plants as they evolved, or do new feeding strategies appear coincident with new plant groups? In particular, in the extensive evidence of herbivore damage present in angiosperm leaf fossils due to newly emerged herbivore groups, do you see diversification of feeding evidence coincidental with diversification of angiosperms? Did anything significant change in the Palaeocene 'between' dinosaur and mammalian herbivore browsers?

W. G. CHALONER. We are pleased to be asked these questions by Dr Collinson, who is probably more competent to offer answers than we are ourselves. We do not have sufficiently tight time control on the first appearance either of leaf-damage types or of the animals known to be involved, to be able to answer these very pertinent questions. If, as Brown & Lawton (this symposium) suggest, the serrations of leaf margins are a defensive adaptation in response to insect marginal feeding, it becomes of particular interest to note the rarity of such non-entire leaves in pre-angiosperm Mesozoic floras, and their near-absence in Palaeozoic leaves.

Phil. Trans. R. Soc. Lond. B (1991)

Fossil evidence for the evolution of biotic pollination

WILLIAM L. CREPET[1], ELSE MARIE FRIIS[2] AND KEVIN C. NIXON[1]

[1] *L. H. Bailey Hortorium, Cornell University, Ithaca, New York 14853–4301, U.S.A.*
[2] *Department of Palaeobotany, Swedish Museum of Natural History, Box 50007, S-104 05 Stockholm, Sweden*

SUMMARY

The Cretaceous–Paleogene history of plants and insects reveals a discernible pattern in the evolution of floral character complexes and insects. Earliest Cretaceous flowers were small apetalous magnoliids with few parts. They co-occurred with a greater variety of anthophilous insects than has previously been supposed, and the idea that Coleoptera were the principal early insect pollinators is in need of review. By the mid-Cretaceous rosid flowers are known with well-developed corollas and the Rosidae are diverse by the late Cretaceous. The more derived asterid floral types are not firmly established until the Tertiary. Nectaries are present in many of the late Cretaceous rosids and may signal the beginning of the most significant evolutionary interaction between Hymenoptera and angiosperms. Advanced floral types in Maastrichtian and early Tertiary deposits are consistent with the appearance of meliponine Apideae (Stingless honeybees) in the late Cretaceous.

1. INTRODUCTION

The history of floral structure and insect pollination are now within reach of paleobotanists. Fossil data are particularly useful in the context of the distribution of insect pollination mechanisms in angiosperms and in the related groups, Bennettitales and Gnetales. It is most likely that angiosperms arose at a time when insect pollination was already established in sister clades. Parallelism and convergence may have been significant in the characters held in common between early angiosperms and some related non-angiospermous groups, but can be fully assessed only with careful phylogenetic analyses.

After the origin of angiosperms, the ongoing relation between angiosperms and insects undoubtedly affected angiosperm radiation and success. It is important to consider how phylogenetically informative the distribution of pollination syndromes in the angiosperms might be and the ecological or evolutionary implications that follow from this distribution.

Our present goal is to update fossil evidence on the evolution of insect pollination mechanisms, and examine in a preliminary way the possible utility and implications of these data (e.g. does fossil evidence reveal patterns relevant to insect pollination and angiosperm phylogeny, and success?). To employ fossil evidence in addressing these questions, certain assumptions are necessary about the relation between structure and mode of pollination and the phylogenetic value of fossils. We have previously considered the historical development and timing of appearance of relevant fossil structures (Crepet 1985; Crepet & Friis 1987; Friis & Crepet 1987; Friis & Endress 1990). However, the relation between fossils and phylogenetic analysis has been the subject of some controversy since the publication of what we regard as a widely mis-

interpreted paper by Patterson (1981). Even so, that paper has served as a focus for several subsequent expositions on the value of fossil evidence in the rigorous context of phylogeny reconstruction (Donoghue *et al.* 1989).

2. CONCEPTUAL FRAMEWORK
(a) *Anthophyte phylogeny*

A phylogenetic framework is necessary for the analysis of the fossil record and, ultimately, for the analysis of the distribution of pollination mechanisms in angiosperms and related groups. Cladistics has emerged as an essential tool for the understanding of phylogeny and for the testing of particular adaptational hypotheses (Coddington 1988). Parsimony analyses provide independent tests of hypothesized character transformations considered to be indicative of particular pollination syndromes.

In recent years a consensus has emerged among botanists that the Gnetales are among the closest living relatives of angiosperms, whereas Bennettitales are the closest fossil group to both angiosperms and Gnetales (Crane 1985; Doyle & Donoghue 1986). At the present time, the evidence for the exact relations of these three groups is somewhat equivocal, and we take here a conservative interpretation in which Gnetales, angiosperms and Bennettitales form an unresolved trichotomy that is probably derived from a Mesozoic seed fern group. Characters that support the monophyletic nature of the angiosperm–Gnetales–Bennettitales ('anthophyte') group include close arrangement of male and female organs (bisexuality), the second integument of the ovule, and possibly granular–columellate pollen wall structure. Numerous features of the extant Gnetales, which are not observable in fossil Bennettitales, can also be cited, and with more

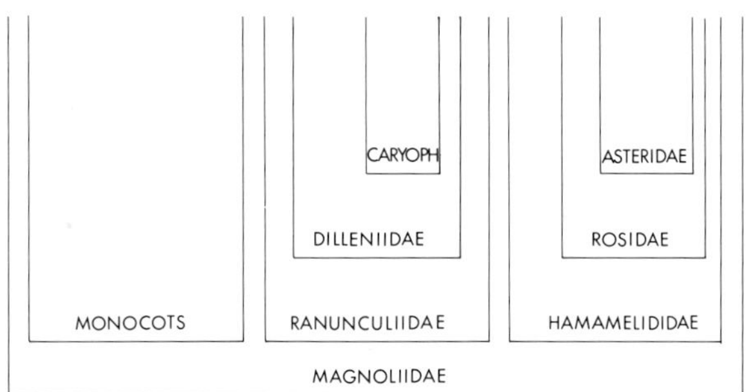

Figure 1. Generalized hierarchic relations of the angiosperm subclasses. This should not be interpreted as a final phylogeny, but only as a preliminary assessment based on very broad character distributions. Note that two major tricolpate/tricolpate-derived pollen-bearing lineages (Ranunculiidae, Hamamelididae) are derived separately from Magnoliidae in this hypothesis.

careful analyses, we anticipate that Gnetales, or some portion of Gnetales, may be found to be closer to the angiosperms than is Bennettitales. The inclusion of *Pentoxylon* in the 'anthophyte' clade as suggested by Crane (1985) and Doyle & Donoghue (1986) is based largely on missing values (Nixon & Davis 1991), as well as characters that reverse on the *Pentoxylon* clade. Thus, our position is that it is premature to include *Pentoxylon* and in any case so little is known about important features of this fossil taxon that it is of little interest to our discussion.

A phylogeny within the angiosperms, particularly one that outlined relations of major groups, would be desirable in the context of the analysis presented below. Unfortunately, the scope of such an analysis, character variation within the angiosperms, lack of equivalent data among the many groups, and the relatively recent emphasis on cladistic methodology applied to angiosperms, have combined to delay such analyses. In the context of this paper, we therefore must utilize a very general tentative hierarchy reflective of some established thoughts on relations among major angiosperm groups.

Based on outgroup comparison, information from published cladograms and classical phylogenetic studies, it is likely that the ancestral angiosperms had a combination of features of one or more of the modern families placed in the subclass Magnoliidae. These ancestral forms most likely had monocolpate pollen, simple pinnately-veined leaves, fleshy stamens with well-developed connective extensions, dithecal tetrasporic anthers, bitegmic ovules, and ethereal oil cells at least in the leaves. The Magnoliidae is clearly a paraphyletic assemblage within which the higher groups of angiosperms are nested, including 'higher' dicotyledons with tricolpate and tricolpate-derived pollen, as well as the basically monocolpate monocotyledons (figure 1). The evidence seems insufficient at this time to determine if the tricolpate and tricolpate-derived angiosperms are a monophyletic group. The prevailing thought recognizes at least two major lines in which the tricolpate condition arose independently from monoaperturate magnoliid ancestors, the ranunculid–dilleniid line, and the hamamelidid–rosid–asterid line (Cronquist 1981; figure 1).

Evidence supports the interpretation that the Asteridae are nested within the Rosidae, and the Rosidae in turn are nested within the Hamamelididae. Thus, these latter groups represent paraphyletic assemblages that have generalized trends of characters, and will most likely not survive careful cladistic analyses.

(b) Angiosperm sister groups

(i) Gnetales

The Gnetales first appear in the Triassic. They are reported throughout the Mesozoic and Caenozoic with a maximum distribution in the early and mid-Cretaceous. They are best represented in the fossil record by dispersed pollen similar to *Ephedra*, although megafossils are rare. Such pollen has been reported *in situ* in small ovulate structures from the Early Cretaceous (Krassilov 1986). Although not fully understood, these structures support a gnetalean affinity of this pollen type.

Among characters Gnetales share with angiosperms are close association of ovulate and staminate organs. This is particularly well developed in *Welwitschia* in which staminate flowers have a central aborted ovule. Pollination is poorly understood in these taxa with the exception of *Ephedra*. *Ephedra* has one species (*E. aphylla*) which is both insect pollinated and anemophilous (Meeuse *et al.* 1990; Niklas & Buchmann 1987).

(ii) Bennettitales

The Bennettitales are remarkable for their superficially flower-like reproductive structures. Both bisexual and unisexual reproductive structures have been reported in early Bennettitales. Dispersed ovulate receptacles and androecia provide primary evidence that some taxa had unisexual reproductive structures, but structure and reproductive biology of these fossils are not well understood. Bennettitales show more variability in microsporophyll structure than in morphology of the ovules or ovulate receptacles. In unequivocally bisexual taxa, conical to dome-shaped multi-ovulate receptacles are subtended by whorled microsporophylls, that are in turn subtended by spirally arranged bracts. Variability in microsporophyll structure through time has been ascribed to a trend reflective of increasing predation or specialization for beetle pollination during the Mesozoic (Crepet 1974; Gottsberger 1988). The possibility that an observed trend within the fossil history of a group might reflect a transformation series related to increasing interactions with beetle pollinators may come to be better evaluated in the context of a phylogenetic analysis of variation in Bennettitales and through applying the precepts of Coddington (1988). In any case, the bisporangiate condition, universality of ovule protecting interseminal scales, direct evidence of beetle-chewing damage of microsporophylls and ovules in *Cycadeoidea*, and highly modified and variable microsporophylls in conjunction with cone position and degree of exposure are all highly suggestive of insect pollination. In *Williamsoniella*, the fleshy micro-

Phil. Trans. R. Soc. Lond. B (1991)

sporophylls are highly reduced with only four to six synangia, and may have served as attractants or rewards.

3. FOSSIL ANGIOSPERMS

(a) *Fossil record*

The earliest evidence of angiosperms in the geological record is monocolpate pollen from Lower Cretaceous (Hauterivian) strata (Brenner 1984). In Hauterivian–Barremian assemblages angiosperms typically comprise less than 1% of the total plant assemblage (see, for example, Crane (1987)). Palynological data and evidence from fossil leaves indicate that the angiosperms underwent a major radiation during the later part of the early Cretaceous (Aptian–Albian) and in the course of a few million years, the group was well established at a global scale.

Cretaceous floras discovered over the past few years include extremely well preserved flowers and have provided a greater range of characters from which to infer both the relations and reproductive biology in angiosperms from this critical time interval. A summary of fossil flowers was given by Crepet & Friis (1987), Friis & Crepet (1987) and Friis & Endress (1990) and stamen features through the Cretaceous were summarized by Friis *et al.* (1991) and we refer the reader to these papers for more detailed descriptions and literature citations.

(b) *Characters of early Cretaceous flowers*

Although data available from dispersed pollen, leaf fossils and floral structures indicate a major radiation of angiosperms during the Aptian–Albian, these floral structures are rather uniform in general organization, and within the dicotyledons only magnoliid and hamamelidid lineages have been identified. An early establishment of rosids as indicated by leaf fossils remains to be documented by floral evidence. The presence of monocotyledons is suggested by dispersed pollen, but a more precise systematic affilation within the monocots has yet to be established for these grains. The presence of magnoliids is documented mainly by a variety of monocolpate, reticulate pollen referred to *Clavatipollenites* and related genera. Although some *Clavatipollenites* show great similarity to pollen of modern Chloranthaceae, wall ultrastructure indicates this complex may represent several distinct magnoliid families (Pedersen *et al.* 1991). A Barremian fossil from Australia with fruits and leaves preserved is described as having a mosaic of chloranthoid and piperalean characters (Taylor & Hickey 1990). A small androecium from the Albian of North America also suggests the presence of the Chloranthaceae (Crane *et al.* 1989). Winteraceous pollen tetrads have been reported from the Aptian–Albian (Walker *et al.* 1983). The Hamamelididae is represented by several kinds of pistillate and staminate platanoid inflorescences (Friis *et al.* 1988) and small unisexual flowers of possible trochodendralean–buxoid affinity (Drinnan *et al.* 1990). *In situ* pollen in these hamamelididean flowers is similar to dispersed pollen assigned to *Tricolpites* and

Retimonocolpites, both of which have an extensive record in the early Cretaceous palynofloras.

Size: fossil flowers are typically minute (0.5–4 mm) and borne in dense inflorescences.

Distribution of sex: flowers are unisexual and arranged in monoclinous or diclinous inflorescences or more rarely bisexual.

Number of floral parts is typically low and there is only one record of a reproductive structure with many parts. The angiospermous affinity of this fossil, however, is ambiguous.

Perianth parts are free, consisting of small membranous or coriaceous tepals. There is no indication that corolla or showy bracts were present in any of the Early Cretaceous forms.

Androecium: stamens are typically fleshy with weakly differentiated filament and anther. Connective tissue between the pollen sacs is extensive and typically expanded beyond the pollen sacs. Anthers are dithecal and tetrasporangiate with relatively small pollen. Dehiscence is valvate or more rarely by longitudinal slits. Dispersed pollen and pollen grains observed in stamens are small, monocolpate or tricolpate–tricolporate with columellate, reticulate exine. A pollenkitt-like substance has been observed in a single taxon.

Gynoecium: carpels are typically free with sessile and sometimes strongly decurrent stigmatic surface, rarely syncarpous. The adaxial suture in some apocarpous forms is apparently incompletely sealed.

Position of parts: an undescribed flower from Portugal has an epigynous perianth whereas all other angiosperm flowers so far identified are completely hypogynous.

Floral symmetry is actinomorphic.

Attractants and rewards: none of the flowers have any indication of a corolla. Attraction of pollinators was most likely by the androecium as in modern magnoliid flowers with similar anthers (Endress 1990). Thus, connective tissue may have been brightly coloured and odoriferous with yellow pollen sacs. Available evidence suggests that the only major reward was pollen. Carpels in pistillate platanoid flowers show superficial resemblance to stamens from corresponding staminate flowers in having an extended apical portion, and may have had attractive function as in modern *Myristica insipida* (Armstrong & Irvine 1989).

(c) *Characters of mid-Cretaceous flowers*

The Cenomanian record of angiosperms documents a further diversification of magnoliid and hamamelidid taxa (Dilcher 1979), and the establishment of several rosid lineages. The major diversification of the rosids, however, did not appear until later in the Cretaceous. Within the Magnoliidae, Lauraceae and Magnoliales have their first appearance. Small fruits with *Clavatipollenites*-type pollen share several characters with fruits of modern Chloranthaceae but differ by their anatropous ovules (Pedersen *et al.* 1991). Floral structures of Hamamelididae show little innovation, but dispersed pollen of the extinct *Normapolles*-complex records a major radiation by the mid-Cenomanian.

Size: fossil flowers are typically small as seen in

earlier flowers, but larger floral structures, about 1–2 cm in diameter, have also been reported, as well as larger fruiting structures.

Distribution of sex: unisexual flowers remain common, but bisexual flowers are also diverse.

Number of floral parts: although number of parts was generally low in early Cretaceous flowers several multiparted taxa are present in the Cenomanian. Most flowers, however, have few parts, and lineages having flowers with trimerous, tetramerous or pentamerous arrangements are known.

Perianth parts are free and may be differentiated into calyx and corolla, although forms with undifferentiated perianths remain common.

Androecium: stamens with only slightly differentiated filaments and anthers, massive connective tissue and apical expansions of the connective are common, but several taxa show distinct filament and anthers with little connective tissue developed. These types typically dehisce by longitudinal slits whereas valvate dehiscence prevails in the fleshy stamens. Staminodea and lateral appendages at the base of the stamens are present in a lauraceous flower (Drinnan *et al.* 1991). Monocolpate and tricolpate–tricolporate pollen are diverse and include reticulate forms as well as forms with continous tectum.

Gynoecium: several taxa with syncarpous ovaries suggest that fusion of carpels was established and diverse.

Position of parts: most flowers are hypogynous, but epigynous taxa are also present.

Floral symmetry is mostly actinomorphic and flowers are shallow and open.

Attractants and rewards: the androecium continues to play a role as attractant in many magnoliids and hamamelidids in which perianth is undifferentiated and pollen was the main reward. Modifications occur in the lauraceous flower in which the lateral stamen appendages probably functioned as food bodies. In several rosids the attractive function had shifted to the corolla. Although nectaries have been recorded in a single flower this is poorly documented and pollination rewards may still have been mainly pollen.

(d) *Characters of late Cretaceous flowers*

The late Cretaceous record of fossil flowers documents a major radiation particularly within the rosids, dilleniids and higher hamamelidids (excluding the Asteridae, figure 1). The latter includes the diversification of the *Normapolles*-complex also documented by an extensive record of fossil pollen. Within the rosids, several lineages of saxifragalean and myrtalean affinities are known. There is also increased diversity within the monocotyledons with flowers, fruits and seeds related to the Liliidae and Arecidae.

Size: flowers are typically very small and larger reproductive organs are preserved in fruiting stage only.

Distribution of sex: most of the flowers are bisexual. Unisexual flowers prevail in the hamamelididean taxa although both bisexual and unisexual flowers have been identified within the Normapolles-complex.

Number of floral parts is typically low and arrangement of parts mostly cyclic.

Perianth is well differentiated into a calyx and a corolla in most flowers. Parts are typically free, but fusion of petals into a short tube has been documented for a single taxon.

Androecium: stamens typically have clearly differentiated filament and anther with weakly developed connective tissue. Pollen is typically small and often clumped together by a pollenkitt-like substance.

Gynoecium: syncarpous gynoecia with well-developed stylar tissue prevail.

Position of parts: there is a remarkably high proportion of epigynous forms. This was reported by Crepet & Friis (1987) and has been further supported by new findings in Portugal. In one flora relative position of perianth and gynoecium could be established in 15 different taxa and in 12 of these, gynoecium is inferior.

Nectaries are present in a variety of taxa typically formed as a disk inserted between androecium and gynoecium. Septacular nectaries were reported in a monocotyledonous taxon from Japan.

Floral symmetry is typically actinomorphic, but the presence of bisymmetrical and zygomorphic forms is documented by floral evidence and implied by indirect evidence from other organs.

Attractants and rewards: the main attractive feature apparently was the corolla. Nectaries are common in many different taxa and nectar had probably become an important reward in the rosid lineages.

(e) *Characters of early Tertiary flowers*

An extensive innovation in floral organization took place during the early part of the Tertiary period and families representing all major subclasses of angiosperm have been identified. In addition to the taxa that persist after the Cretaceous, there are several major groups of taxa that radiate in the early Tertiary, including the legumes and Euphorbiaceae. Higher Hamamelididae and grasses appear early in the Tertiary or in the uppermost Cretaceous and radiate in the Tertiary.

Size: direct evidence from fossil flowers and indirect evidence from other organs indicate that there is a considerable size range in the early Tertiary flowers.

Distribution of sex: bisexual flowers prevail and this is reflected in the more modern flora of the early Tertiary. Higher Hamamelididae are very well represented as are other wind pollinated dicots.

Number of floral parts is typically low and arranged in whorls of 3, 4 or 5. Evidence from fossil fruits and seeds indicates that an increase in the number of stamens took place in several separate lineages.

Perianth parts are typically differentiated into calyx and corolla. Sepals are mostly free. Petals are free or laterally fused to form funnel-shaped or wider corolla tubes.

Androecium: flowers exhibit a wider range of stamen morphology than seen in the Cretaceous although forms with distinct filaments and simple tetrasporangiate anthers are the most common forms.

Exserted brush like androecia occur in several taxa and stamen arrangements exhibit various modifications that have not been seen in earlier fossils (e.g. bilaterally arranged partially-fused stamens of papilionoid legumes).

Gynoecium: carpels are typically fused to form syncarpous ovaries with distinct styles. Fruits vary widely in morphology. Fleshy fruits as well as various winged fruits were diverse, indicating more specialized dispersal systems than seen in the Cretaceous angiosperms.

Position of parts: epigynous flowers are present but less common than hypogynous types.

Floral symmetry: the Tertiary record of flowers provides the earliest floral evidence of strongly zygomorphic types formed by differentiation of corolla lobes. Indirect fossil evidence indicates that zygomorphy was established in several lineages.

Attractants and rewards: the attractive function was mainly by the corolla, or in some brush flowers, the attractive function was probably by the stamens. Stamens in these flowers were not massive as in many Cretaceous forms, but protruding from the flowers by long thin filaments. Reward was most likely nectar and to some extent pollen.

4. INSECT FOSSIL RECORD

There have been interesting developments in our knowledge of the fossil record of anthophilous insects since we last reviewed the history of insect pollination (Crepet & Friis 1987). In that review, we predicted the presence of bees in the late Cretaceous based on the fossil record of angiosperm floral characters and families with particularly distinctive combinations of characters. Meliponinae have subsequently been reported from the late Cretaceous of New Jersey (Michener & Grimaldi 1988). The records of other orders of anthophilous insects have improved and some spectacular new fossil insect localities have been reported and their faunas described.

(a) *Coleoptera*

Beetles are well established in the Permian with several flower-visiting orders present by the Jurassic (including Nititulidae, Elateridae and Buprestidae). They continue to diversify in the Mesozoic with many families associated with pollen- and nectar-feeding present throughout that era (Handlirsch 1906–1908; Carpenter 1976; Crowson 1981). Phylogenetic relations among certain 'primitive' groups of beetles (Boganiidae (modern pollen feeders on cycads and shrubby angiosperms where cycads do not occur; Endrody-Younga & Crowson 1986), Hobartiinae and others), suggest an interesting relation between mycophagy and pollen feeding early on in Coleoptera (Wheeler, personal communication). Recent reports include Staphylinidae with both pollen and nectar feeders from the early Jurassic and late Cretaceous, and Chrysomelidae (leaf chewing beetles) that are typically pollen feeders. Beetle-like borings have been reported in the reproductive structures of Cretaceous *Cycadeoidea* (Crepet 1974).

In addition to the highly derived syndromes (Gottsberger 1988) associated with complex large-flowered Magnoliidae, beetles may pollinate small flowers with exposed nectar or with nectar shallowly hidden in addition to pollen. Some exceptional beetles are well suited to nectar feeding, with modified maxillae up to 12 mm long (Proctor & Yeo 1973).

(b) *Diptera*

Flower visiting Diptera are distributed among all three suborders of flies: Nematocera (usually considered the most primitive), Brachycera and Cyclorrhapha. The fossil record of families that include flower-visiting flies is good and includes a significant number of new discoveries (Grimaldi 1990). In Nematocera there have been additional reports of Cretaceous Tipulidae, a family that extends to the Jurassic and pollinates Saxifragaceae today (Raynder & Waters 1990). Bibionidae have been reported throughout the Cretaceous, Mycetophilidae from Jurassic through Cretaceous and Chironomids from the early Cretaceous. The Ceratopogonidae are now known from the late Cretaceous and Pycodidae from the early Cretaceous.

In Brachycera, Rhagioniidae have been reported from the early Jurassic, but this has been regarded as insufficiently substantiated (Grimaldi 1990). Empididae have a suggested Jurassic origin and more reliable post mid-Cretaceous fossil record and are thought to be primtively predatory with specialization for nectar feeding following angiosperm radiation. The earliest known fossil empids appear to be flower visitors (Negrobov 1978). Asilidae and possible ancient sister groups are known to the Jurassic. Bombyliidae have a much improved fossil record with three new genera each with a relatively short proboscis known from the early Cretaceous (Zaitzev 1986), suggesting an origin of the family in the Jurassic (Grimaldi 1990). There have been recent reliable reports of Phoridae from the early Cretaceous leading to the suggestion that they originated much earlier (Grimaldi 1989).

Cyclorrhapha Syrphidae have a poor fossil record, but it has been suggested that they were diverse by the Tertiary.

Nematoceran pollinators typically pollinate flowers that are relatively small with exposed nectar. There are some taxa that pollinate flowers with small tubular flowers or with partially concealed nectar. Brachyceran families pollinate a wider variety of floral types and sizes from small flowers to large flowers; from those with exposed nectaries to concealed nectar (unusual), from small tubular to large tubular flowers.

(c) *Lepidoptera*

There have been more reports of fossil Lepidoptera since our last review, but few of them offer direct fossil evidence that alter our present understanding of timing in lepidopteran evolution. The earliest reports of Lepidoptera are from the Jurassic but Lepidoptera are thought to have diverged from their sister group, the Trichoptera, in the Permian. New records of micropterygid moths (pollen chewing) have been reported from the early Cretaceous and fossils have now also been discovered in the southern Hemisphere (Martens-Neto & Vulcano 1989). A new heteroneuran (Heteroneura are nectar feeders) moth has been reported from the early Cretaceous (Skalski 1984). If this is accurate, it may represent the first generally accepted evidence of the group. Whalley (1986) questioned the nature of Campanian eggs of Noctuidae (Gall & Tiffney 1983) and recognizes no Noctuidae earlier than the Miocene. Recent reports and analyses on the whole suggest an earlier origin for Lepidoptera, including butterflies, than is presently reflected by the fossil record. Lepidoptera mines on seed fern foliage of the late Jurassic are reported as suggestive of Nepticulidae (Grogan & Szadziewski 1988), but they are poorly preserved and must be interpreted with some caution. Phylogenetic analysis of Papilioninae suggest a mid-Cretaceous origin for some groups (Miller 1987). Even Whalley (1986), suggests that Heteroneura must be missing from the Cretaceous fossil record. Unequivocal butterflies appear during the Lower Tertiary (Papilion-ideae, Nymphalidae, Satryidae, Lycaenidae) and are well represented by the Eocene. The moth superfamilies Gelechioidea, Cossoidea, Pyraloidea, Geometroidea and Copromorphoidea are present in Eocene–Oligocene deposits.

(d) Hymenoptera

There have been important discoveries of fossil Hymenoptera since our last review. Symphyta are well known from The Triassic on and are well represented in Cretaceous sediments. Cretaceous Symphyta have even been found preserved with pollen of seed ferns in their guts (Krassilov & Rasnitsyn 1982). There has been a dramatic increase in the number of reports of Apocrita in the past five years (see, for example, Darling & Sharkey (1990)). Among the parasitic wasps, there have been Cretaceous reports of ichneumonids and Cynipidae. There have been numerous new and well-founded reports of Aculeata from the Cretaceous. Chrysididae now have a Cretaceous record and scoliids and pompilids are also known from the Cretaceous. Vespoids appear in the Lower Cretaceous (cf. Carpenter & Rasnitsyn 1990). Specidae, the paraphyletic stem group of bees, are now well known from the early Cretaceous. The most dramatic fossil discovery in recent years is the discovery of Apidae (*Trigona*, Meliponini) from the late Cretaceous (Michener & Grimaldi 1988).

5. MAJOR STAGES IN THE EVOLUTION OF INSECT POLLINATION

There is a discernible pattern in the evolution of floral structure and insect pollination based on the fossil history of plants and insects. In the anthophytes, Bennettitales and Gnetales appear before the flowering plants and overlp with them chronologically. The Bennettitales may be interpreted on the basis of several kinds of evidence as having been pollinated by beetles (Crepet 1974), and flies and Hymenoptera cannot be ruled out in certain taxa. We interpret Gnetales as basally bisexual (with close association of staminate and pistillate organs) and infer that insect pollination is basal in the group. Insect pollination was in place at the time of angiosperm origin and in early angiosperms the androecium probably served as the sole reward and attraction for pollinators as in Bennettitales. There is even parallelism (or homology pending further analysis) between the fleshy microsporophylls of *Williamsoniella* and those of early angiosperms.

The Cretaceous and early Tertiary fossil record is broadly consistent with the generalized hierarchical relations of the major angiosperm groups as presented in figure 1, and suggests that the earliest angiosperms may have been small-flowered, apetalous magnoliids with few parts, either asymmetric or cyclically arranged. By the mid-Cretaceous, rosid flowers are known, as well as larger multiparted magnoliids. The more derived asterid floral types are not firmly established until the Tertiary, when numerous modern families appear. This pattern suggests that the hamamelidid–rosid–asterid line is probably a very early offshoot of small-flowered magnoliids, and that the larger-flowered magnoliids, long considered indicative of the most primitive angiosperms, may mostly belong to later lineages that might be considered evolutionary cul-de-sacs.

Earliest Cretaceous floral types co-occurred with a greater variety of anthophilous insects than has previously been supposed. Representatives of the four 'major orders' of anthophilous insects present in the early Cretaceous now pollinate a variety of simple, usually small floral types in mutli-flowered inflorescences with exposed rewards. Thus, the idea that Coleoptera were the earliest insect pollinators is in need of review because the efficacy of other anthophilous insect groups present at the time in the context of their contemporary angiosperms is either proven (cf. references) or cannot be dismissed.

The early Cretaceous (Aptian–Albian) angiosperms are poorly represented by floral structures, but remains of Chloranthaceae, Trochodendrales–buxoid fossils and platanoids are present in Albian deposits. Older pollen of chloranthaceous and winteraceous affinity suggests early pollination by beetles, flies and micropterygids consistent with the insect fossil record. Some chloranthoids may have been wind pollinated, based on dispersed pollen similar to *Ascarina*, but in the absence of overall floral structure, such interpretations should be made with caution. Cenomanian fruits with abundant *Clavatipollenites* pollen adhering to the sessile stigma indicate that at least some chloranthoids were insect pollinated.

Meristic stabilization of floral parts in cycles of threes (Lauraceae) or four to five (Rosidae) occurs by the Cenomanian. Large meristically indeterminate flowers (of Magnoliales affinity) appear at the same time. The first specialization for beetle pollination in angiosperms is represented in the Cenomanian Magnoliales. Lauraceae are frequently pollinated by bees today, but many of these reports are somewhat equivocal, as they involve *Apis mellifera* pollination of cultivated species (House 1989). However, flies and beetles have been documented as pollinating Lauraceae in their natural setting. Thus, the occurrence of Lauraceae cannot be considered strong evidence for earliest bee pollination.

According to Darling & Sharkey (1990) apparent stasis in proportions of higher level taxa in Hymenoptera throughout the Mesozoic implies a disconnection between angiosperm and hymenopteran radiations. This may be a premature conclusion, because it is not the only explanation of observed patterns in Hymenoptera in the fossil record, nor does it take into account what is known about the evolution of floral structure in the Cretaceous. From the fossil evidence of the Early Cretaceous it seems apparent that there was a general similarity in attractant and reward structures in angiosperms until the Cenomanian specialization for beetle pollination in Magnoliales and the appearance of the corolla as an attractant in Lauraceae and Rosidae. Although stabilization of cyclical 4-5-merous floral parts in the latter set the stage for the evolution of the floral tube and other corolla characters now associated with pollination in Apidae (and Lepidoptera), no specializations that can be associated with pollination by Apidae appeared before Santonian–Campanian flowers with corolla tubes. Nor do pollinators in the early Cretaceous–Campanian interval exhibit features or taxonomic affiliations that suggst any breakthroughs in attractants or rewards on the parts of the plants. It is the Cenomanian that marks the appearance of the Rosidae with its associated character complex related to pollination by generally more advanced insect pollinators. In the late Cretaceous Santonian–Campanian discoveries of numerous rosid taxa with well-developed nectaries and well-developed corollas accompany the specimens with fused corolla tubes and may signal the beginning of the most significant evolutionary interaction between Hymenoptera and angiosperms. The proliferation of advanced floral types in the Maastrichtian and Early Tertiary is consistent with this possibility as is the dramatic appearance of meliponine Apidae in the Late Cretaceous. Thus, the interpretation of apparent stasis in the hymenopteran line in the face of angiosperm radiation may be due to failure to recognize the strong relation between the character complex embodied in early rosids and the Hymenoptera. It is important to point out that these late Cretaceous rosids were possibly pollinated by other groups of insects as well because a high proportion of them are epigynous, a condition Grant (1950) associates with beetle pollination, but this hypothesized correlation invites confirmation based on more extensive data on pollination in extant epigynous flowers.

The Paleogene signals the modernization of angiosperm families and reveals a proliferation of floral types with highly adapted corollas as well as families whose radiations have been intimately associated with bees (Leguminosae, Euphorbiaceae) or which have highly modified perianths as attractants or novel rewards (e.g. elaiophores in Eocene Malphigiaceae). Haustellate Lepidoptera including diverse moths and butterflies appear in the Paleogene. The fossil evidence at this time illustrates an uppermost Cretaceous–Paleogene radiation of highly modified corollas associated with pollinators (bees) that by correlation and mechanism (fidelity as a potential isolating mechanism, efficiency, etc.) are thought to have contributed greatly to diversification in angiosperms. This is also a time of the greatest rate of appearance of new angiosperm taxa. We suggest that the fossil record as now understood reflects a rapid late Cretaceous radiation of bees and the types of flowers and families associated with them that extends well into the Paleogene. The higher Lepidoptera and higher Diptera (Cyclorrhapha) and Coleoptera (Chrysomelidae) have a similar temporal relation with the angiosperms. It is also possible that the relation between the apparent time of appearance of bee pollination and rate of angiosperm diversification reflects a causal relation of some dimension. This is plausible based on available evidence but is a hypothesis that can be tested only with the greatest difficulty.

REFERENCES

Armstrong, J. E. & Irvine, A. K. 1989 Floral biology of *Myristica insipida* Myristicaceae: a distinctive pollination syndrome. *Am. J. Bot.* **76**, 86–94.

Brenner, G. J. 1984 Late Hauterivian angiosperm pollen from the Helez Formation, Israel. *Abstracts, 6th Int. Palynol. Conf., Calgary* p. 15.

Carpenter, F. M. 1976 Geological history and the evolution of the insects. *Proceed. 15th Congress Entomol.* (ed. D. White), pp. 63–70. Washington, DC: American Entomological Society.

Carpenter, J. M. & Rasnitsyn, a. P. 1990 Mesozoic Vespidae. *Psyche* **97**, 1–20.

Coddington, J. A. 1988 Cladistic tests of adaptational hypotheses. *Cladistics* **4**, 3–22.

Crane, P. R. 1985 Phylogenetic analysis of seed plants and the origin of angiosperms. *Ann. Missouri Bot. Gard.* **72**, 716–793.

Crane, P. R. 1987 Vegetational consequences of the angiosperm diversification. In *The origins of angiosperms and their biological consequences* (ed. E. M. Friis, W. G. Chaloner & P. R. Crane), pp. 107–144. Cambridge University Press.

Crane, P. R., Friis, E. M. & Pedersen, K. R. 1989 Reproductive structure and function in Cretaceous Chloranthaceae. *Plant Syst. Evol.* **165**, 211–226.

Crepet, W. L. 1974 Investigations of North American cycadeoids: the reproductive biology of *Cycadeoidea*. *Palaeontogr.* B **148**, 144–159.

Crepet, W. L. 1985 Advanced (constant) insect pollination mechanisms: patterns of evolution and implications *vis-a-vis* angiosperm diversity. *Ann. Missouri Bot. Gard.* **71**, 607–630.

Crepet, W. L. & Friis, E. M. 1987 The evolution of insect pollination in angiosperms. In *The origins of angiosperms and their biological consequences*. (ed. E. M. Friis, W. G. Chaloner & P. R. Crane), pp. 181–201. Cambridge University Press.

Cronquist, A. 1981 *An integrated system for classification of flowering plants* (1262 pages.) New York: Columbia University Press.

Crowson, R. A. 1981 *The biology of Coleoptera.* (802 pages.) London: Academic Press.

Darling, D. C. & Sharkey, M. J. 1990 Order Hymenoptera. In *Insects from the Santana Formation, Lower Cretaceous, of Brazil* (ed. D. A. Grimaldi) (*Bull. Am. Mus. nat. Hist.* **195**), pp. 123–153.

Dilcher, D. L. 1979 Early angiosperm reproduction: an introductory report. *Rev. Palaeobot. Palynol.* **27**, 291–328.

Donoghue, M. J., Doyle, J., Gauthier, J., Kluge, A. & Rowe, T. 1989 The importance of fossils in phylogenetic reconstructions. *A. Rev. Ecol. Syst.* **20**, 431–460.

Doyle, J. A. & Donoghue, M. J. 1986 Seed plant phylogeny and the origin of angiosperms: an experimental cladistic approach. *Bot. Rev.* **52**, 321–431.

Drinnan, A. N., Crane, P. R., Friis, E. M. & Pedersen, K. R. 1990 Lauraceous flowers from the Potomac Group (mid-Cretaceous) of eastern North America. *Bot. Gaz.* **151**, 370–384.

Drinnan, A. N., Crane, P. R., Pedersen, K. R. & Friis, E. M. 1991 Angiosperm flowers and tricolpate pollen of bux-aceous affinity from the Potomac Group (mid-Cretaceous) of eastern North America. *Am. J. Bot.* **78**, 153–176.

Endress, P. K. 1990 Evolution of reproductive structures and functions in primitive angiosperms (Magnoliidae). *Mem. New York Bot. Gard.* **55**, 5–34.

Endrody-Younga, S. & Crowson, R. A. 1986 Boganiidae, a new beetle family for the African fauna (Coleoptera: Cucujoidea). *Ann. Transvaal Mus.* **34**, 253–273.

Friis, E. M. & Crepet, W. L. 1987 Time of appearance of floral features. In *The origins of angiosperms and their biological consequences* (ed. E. M. Friis, W. G. Chaloner & P. R. Crane), pp. 145–179. Cambridge University Press.

Friis, E. M., Crane, P. R. & Pedersen, K. R. 1988 Reproductive structure of Cretaceous Platanaceae. *Biol. Skr. Dan. Vid. Selsk.* **31**, 1–56.

Friis, E. M., Crane, P. R. & Pedersen, K. R. 1991 Stamen diversity and in situ pollen of Cretaceous angiosperms. In *Pollen and spores: pattern and diversification* (ed. S. Blackmore). Oxford: Clarendon Press. (In the press.)

Friis, E. M. & Endress, P. K. 1990 Origin and evolution of angiosperm flowers. *Adv. Bot. Res.* **17**, 99–162.

Gall, L. F. & Tiffney, B. H. 1983 A fossil noctuid moth egg from the late Cretaceous of Eastern North America. *Science Wash.* **219**, 507–509.

Gottsberger, G. 1988 The reproductive biology of primitive angiosperms. *Taxon* **37**, 630–643.

Grant, V. 1950 The protection of the ovule in flowering plants. *Evolution* **4**, 179–201.

Grimaldi, D. A. 1989 The genus *Metropina*, Diptera, Phoridae from Cretaceous and Tertiary Ambers. *J. N. Y. Ent. Soc.* **97**, 65–72.

Grimaldi, D. A. 1990 Diptera. In *Insects from the Santana Formation, Lower Cretaceous, of Brazil* (ed. D. A. Grimaldi) (*Bull. Am. Mus. Nat. Hist.* **195**), pp. 164–183.

Grogan, W. L. & Szadziewski, R. 1988 A new biting midge from Upper Cretaceous Cenomanian amber from New Jersey, USA. Diptera, Ceratopogonidae. *J. paleontol.* **62**, 808–812.

Handlirsch, A. 1906–1908 *Die Fossilen Insekten und die Phylogenie der Rezenten Formen.* Leipzig: Englemann.

House, S. M. 1989 Pollen movement to flowering canopies of pistillate individuals of three rainforest tree species in tropical Australia. *Aust. J. Ecol.* **14**, 77–94.

Krassilov, V. A. 1986 New floral structure from the Lower Cretaceous of the Lake Baikal area. *Rev. Palaeobot. Palynol.* **47**, 9–16.

Krassilov, V. A. & Rasnitsyn, A. P. 1982 A unique find: pollen in the intestine of Early cretaceous saw flies. *Paleontol. Journ.* **16**, 80–97.

Martins-Neto, R. G. & Vulcano, M. A. 1989 Amphies-menoptera, Trichoptera, plus Lepidoptera in the Santana Formation, Lower Araripe Basin Northeast Brazil I. Lepidoptera, Insecta. *Ann. Acad. Braz. Cienc.* **61**, 459–466.

Meeuse, A. D. J., DeMeijer, A. H., Mohr, O. W. P. & Wellinga, S. M. 1990 Entomophily in the dioecious gym-nosperm *Ephedra aphylla* Forsk. *Is. Journ. Basic & App. Plant Sci.* **39**, 113–124.

Michener, C. D. & Grimaldi, D. A. 1988 A *Trigona* from late Cretaceous amber of New Jersey (Hymenoptera: Apidae: Meliponinae). *Am. Mus. Novit.* **2917**, 1–10.

Miller, J. S. 1987 Phylogenetic studies in the papilioninae, Lepidoptera, Papilionidae. *Bull Am. Mus. nat. Hist.* **186**, 365–512.

Negrobov, O. B. 1978 Flies of the superfamily Empidoidea from the Cretaceous Retinites of Northern Siberia Russian-SFSR-USSR. *Palaeontol. Zh.* **2**, 81–90.

Niklas, K. J. & Buchmann, S. L. 1987 Aerodynamics of pollen capture in two sympatric *Ephedra* species. *Evol. Evol.* **41**, 104–123.

Nixon, K. C. & Davis, J. I. 1991 Polymorphic taxa, mising values, and cladistic analysis. *Cladistics* (In the press).

Patterson, C. 1981 Significance of fossils in determining evolutionary relationships. *A. Rev. Ecol. Syst.* **12**, 195–223.

Pedersen, K. R., Crane, P. R., Drinnan, A. N. & Friis, E. M. 1991 Fruits from the mid-Cretaceous of North America with pollen grains of the *Clavatipollenites* type. *Grana* **29**. (In the press.)

Proctor, M. & Yeo, P. 1973 *The pollination of flowers.* London: Collins.

Rayner, R. J. & Waters, S. B. 1990 A Cretaceous crane fly, Diptera, Tipulidae: 93 Million years of stasis. *Zool. J. Linn. Soc.* **28**, 253–272.

Skalski, A. W. 1984 A new lower Cretaceous Lepidoptera, Homoneura. *Bull. Pol. Acad. Sci. Biol.* **32**, 389–392.

Taylor, D. W. & Hickey, L. J. 1990 An Aptian plant with attached leaves and flowers: implications for angiosperm origin. *Science Wash.* **247**, 702–704.

Walker, J. W., Brenner, G. J. & Walker, A. G. 1983 Winteraceous pollen in the Lower Cretaceous of Israel: early evidence of a magnolialean angiosperm family. *Science, Wash.* **220**, 1273–1275.

Whalley, P. 1986 A review of current fossil evidence of Lepidoptera in the Mesozoic. *Biol. J. Linn. Soc.* **28**, 253–272.

Zaitzev, V. F. 1986 New species of Cretaceous fossil bee flies and a review of paleontological data on Bornbyliidae, Diptera. *Entomol. Obozr.* **65**, 815–825.

Discussion

A. J. LACK (*School of Biological and Molecular Sciences, Oxford Polytechnic, U.K.*). The authors emphasized early insect pollination in the fossils. Many extant Chloranthaceae, however, are said to be wind pollinated. This includes *Ascarina* with pollen very similar to fossil *Clavatipollenites*. Is wind pollination as primitive as insect pollination or are all the earliest plants insect pollinated.

E. M. FRIIS. The only megafossil evidence of the plants that produced pollen of the *Clavatipollenites* type is the small fruits of *Couperites mouldinensis* from the early Cenomanian of eastern North America (Pedersen *et al.* 1991). These fruits have a single anatropous seed and have an elongated, sessile stigmatic surface. Clumps of *Clavatipollenites* pollen have been observed adhering to the stigmatic area in several specimens. Their abundant occurrence may indicate insect pollination rather than wind pollination. SEM and TEM studies of pollen wall ultrastructure document a close resemblance with modern pollen of *Ascarina*. Fruit and seed morphology of *Couperites*, however, indicate that although *Couperites* is more closely related to members of the Chloranthaceae than to any other modern magnoliid family, its anatropous seeds preclude an inclusion in the Chloranthaceae as currently circum-scribed.

E. A. JARZEMBOWSKI (*Booth Museum of Natural History, Brighton, U.K.*). In the search for interactive models, the rapid radiation of angiosperms during the Cretaceous is perhaps best explained by major innovations in insects rather than by postulating general pollination. The fast accumulating evidence of Cretaceous insect life suggests significant evolution in the aculeate Hymenoptera. Generalized aculeates such as Baissodidae are first found in the early Cretaceous (Rasnitsyn 1975) but specialized groups such as stingless bees had already evolved by the late Cretaceous (Michener & Grimaldi 1988); that is, the important switch from prey catching to pollen collecting in wasps occurred during this period (Jarzembowski 1991*a*). There is little evidence of honeydew producing insects and extrafloral nectaries in the early Cretaceous (Jarzembowski 1989) which could have increased the value of floral nectaries as a source of liquid carbohydrate rather than protein. The hypothesis could be tested by studying pubescence in fossil insects.

The fossil record of Lepidoptera is probably poor (with the possible exception of microlepidopterans in amber) because like other scaley insects, Lepidoptera trap air and are less likely to become waterlogged and incorporated in aquatic sediments in an undecayed form (E. A. Jarzembowski, personal observation). In addition, there are problems in distinguishing primitive Lepidoptera from Trichoptera in the fossil record (see comments in Jarzembowski (1991*b*).

References

Jarzembowski, E. A. 1989 A fossil aphid (Insecta: Hemiptera) from the Early Cretaceous of southern England. *Cretaceous Res.* **10**(3), 239–248.

Jarzembowski, E. A. 1991*a* New insects from the Weald Clay of the Weald. *Proc. Geol. Assoc.* **102**. (In the press.)

Jarzembowski, E. A. 1991*b* The geological age of insects. *Lond. Nat.* (In the press.)

Rasnitsyn, A. P. 1975 Vysshie pereponchatokrylye mezozoya. *Trudy Paleontol. Inst.*, **147**, 1–134.

Note added in proof (26 July 1991):
Recently discovered fossils germane to our analysis from a North American Cenomanian locality were not available until this chapter was in press. The novel fossils include hamamelidaceous flowers with staminodal nectaries, staminal tubes and sepal cups. At the same locality are fossils of two dilleniid taxa with fused corollas. These fossils therefore are the oldest examples of certain characters critical to the evolution of advanced insect pollination, and add support to our model for the timing of the relationship between angiosperms and advanced pollinators.

Fossil evidence of interactions between plants and plant-eating mammals

MARGARET E. COLLINSON[1] AND JERRY J. HOOKER[2]

[1] *Division of Biosphere Sciences, King's College London, Campden Hill Road, London W8 7AH, U.K.*
[2] *Palaeontology Department, The Natural History Museum, Cromwell Road, London SW7 5BD, U.K.*

SUMMARY

We document changes in mammalian dietary and foraging locomotor adaptation, and appearances and developments of angiosperm fruiting strategies and vegetation types since the late Cretaceous in the Euramerican region, and to some extent in low latitude Africa. These changes suggest: (i) an expansion in the exploitation of dry fruits and seeds by mammals on the ground as well as in the trees after the terminal Cretaceous dinosaur extinction; (ii) a relation between large nuts and rodents, which appear in the late Palaeocene and radiate in the late Eocene; (iii) a relation between primates and fleshy fruits established in the early–Middle Eocene when tropical forests reached their maximum latitudinal extent; (iv) a hiatus of several million years in the vertebrate exploitation of leaves after dinosaur extinction and before the first few mammalian herbivores in the Middle Palaeocene, followed by an expansion in the late Eocene when climates cooled and more open vegetation became established.

1. INTRODUCTION

A glance at a matrix plotting dietary against foraging locomotor adaptive types shows that nearly all possible combinations are fulfilled by modern mammals (figure 1). Understandable exceptions are scansorial, arboreal and aerial grazers and aquatic frugivores. In the late Cretaceous, however, the pattern was quite different with very few boxes being occupied. Adding the then dominant group of land-based vertebrates, the dinosaurs, fills more but still leaves many gaps. By examining the changes that have taken place in locomotor and dietary adaptations in mammals we can begin to understand the ways in which they interacted with angiosperm-dominated vegetation during the past 70 Ma.

2. SOURCES OF EVIDENCE

(a) *Plant fossils*

Plants are represented in the fossil record by a variety of parts ranging from minute dispersed pollen grains to large *in situ* tree stumps and from isolated fruits, seeds and leaves, to infructescences and foliage organically connected to one another. These provide direct evidence of the nature of plant material available to mammals.

In situ tree stumps with associated litter provide direct evidence of ancient forest vegetation. Even when only isolated organs are available, functional morphology combined with taphonomic and sedimentological evidence permit reconstruction of ancient vegetation. Ancient vegetation and climate can be reconstructed using evidence from leaf and wood physiognomy (Friis *et al.* 1987; Wolfe 1990; Wolfe & Upchurch 1987; Collinson 1990).

(b) *Mammal fossils*

Land-based mammals (marine ones are excluded here) most frequently occur as isolated teeth; jaws and postcranial remains are also common but complete skeletons are rare. Those which occur at the German Middle Eocene site of Messel are often enhanced by fur impressions and gut contents (Schaal & Ziegler 1988). Understanding of mammal–plant interaction requires knowledge of mammalian diet and locomotion. Dominant locomotor adaptation is almost always associated with foraging.

(i) *Locomotion*

Locomotion can be deduced from postcranial remains either from isolated limb bones or, if available, from complete skeletons. For instance, to show arboreality requires evidence of prehensility of hand, foot or tail (Jenkins & Krause 1983). Trunk-climbing ability (scansoriality) is more difficult to ascertain, but is typified by strong laterally compressed claws (Koenigswald in Schaal & Ziegler (1988)) and some mobility of the ankle (Szalay 1984). In contrast, a mammal restricted to the ground ('large' ground mammal (LGM) of Harrison (1962) and Andrews *et al.* (1979), but in effect not always large) is recognized especially by laterally restricted mobility of limb articulations, also often presence of hooves and reduction of digits. The class of 'small' ground mammal (SGM) (more

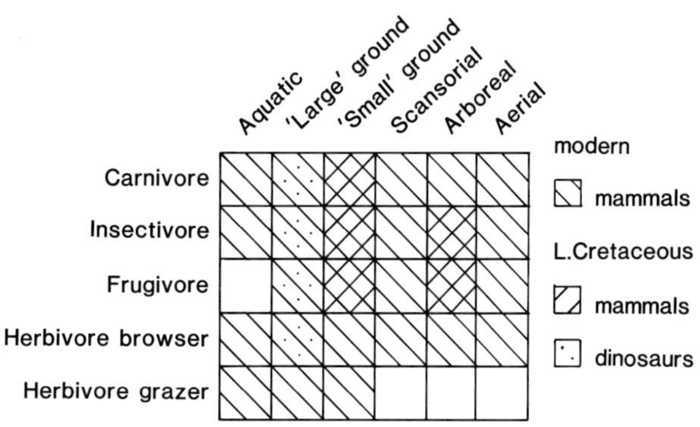

Figure 1. Matrix of dietary and locomotor categories for modern (world) and North American late Cretaceous (see figure 2) land-based mammalian and dinosaur faunas.

appropriately termed semiterrestrial) is probably the most difficult to recognize, as it is intermediate between LGM and scansorial, with the ability to move from the ground into bushes but may also be fossorial (burrowing). P. Andrews (personal communication, 1991) is currently revising Harrison's locomotor classes.

Postcranial information from the nearest relative (nearly always fossil) is used to infer locomotor class for any species with no known postcranials. Usually the relation is no more distant than intrafamilial. Late Cretaceous and Palaeocene taxa are relatively poorly known postcranially. Reliability for locomotor adaptations improves in younger strata.

(ii) *Diet*

Mammalian diet can be inferred from tooth morphology by using modern dental analogues with known diet (see Collinson & Hooker in Friis *et al.* (1987)). Relevant features of teeth are gross morphology, gross wear pattern and microwear, which are apparently interdependent. Mastication occurs in two principal modes: translatory and puncture–crush. Each produces a characteristic pattern of wear (comprehensively reviewed by Janis in Boucot (1990) pp. 241–259; Teaford 1988).

We are here concerned with fossil plant-eating mammals and thus with deducing frugivory (construed here as a diet of fruits and seeds) and herbivory (diet of the green parts of plants, including bark). Herbivory is divided into grazing (feeding on grassland grasses) and browsing (all other types of plant). Typically a herbivorous type of cheek tooth (see, for example, figure 4b) will have little emphasis on puncture–crush but strong cresting with emphasis on translatory movement. Buccal phase shearing facets form as the teeth move towards occlusion, and lingual phase grinding facets form as they move out of occlusion (resulting in distinct wear facets with mainly unidirectional microwear striations). A frugivorous cheek tooth (see, for example, figure 4c) will have low, rounded cusps with little cresting, and emphasis on puncture–crush over translatory movement. This enhances wear at the cusp tips which are dominated by multidirectional microwear striations (Rensberger 1986; Teaford 1988). Intermediate morphologies suggest intermediate (mixed) diets. Details of microwear

may be influenced by variations in hardness of diet, occlusal pressure and size of animal (Janis & Fortelius 1988; Rensberger 1978).

3. MATERIAL

Fossil information is spatially patchy. Our data are largely restricted to the mid- to high latitudes of the Northern Hemisphere. Only those elements of ancient life which were living close enough to depositional settings to be preserved are ever available for study. Assemblages from southern England nevertheless have mammals and plants preserved in the same or adjacent levels (Collinson & Hooker in Friis *et al.* (1987)).

4. EARLY CRETACEOUS AND PALAEOCENE
(a) *Mammals*

The earliest mammals were late Triassic morganucodontids, kuehneotheriids and haramiyids (Lillegraven *et al.* 1979). These were small, mostly insectivorous types though the last probably included fruit in their diet. The multituberculates, the diet of which had a significant plant (fruit) component, first appeared in the late Jurassic and became abundant and diverse in the late Cretaceous.

Postcranial evidence is unknown for most Mesozoic mammals but nearly all late Cretaceous multituberculates with known postcranial remains were arboreal (Jenkins & Krause 1983). They possessed a prehensile tail, opposable hind toe and mobile ankle joint.

The teeth of ptilodontoid multituberculates include a large, ridged blade-shaped premolar in the lower jaw biting against a grooved, cuspate upper tooth. The nearest modern dental analogue is *Burramys* (mountain possum) which has a similar blade in both upper and lower premolars. *Burramys* holds hard-shelled seeds and insects with hard cuticles at the side of the mouth and cracks them with blade-like premolars extracting contents with forceps-like incisors. In addition, it takes dry seeds from opened capsules and tackles drupes by using lower incisors to dig through flesh and then crack the stone (Kerle 1984).

Ptilodontoids are thought to have chewed in a comparable way using slicing–crushing action (Krause 1982) to deal with a diet of mainly hard and dry seeds and fruits. The generally larger taeniolabidoid multituberculates lacked the blade-like premolar but had gnawing, although rooted, incisors and were probably dominantly frugivorous.

The only other group of plant-eating mammals in the late Cretaceous was the didelphoid marsupials (opossums). These also appear to have eaten insects and fruit but were much less specialized for their fruit eating, having teeth of insectivore type merely with less cresting. (One genus with blunter cusps may have fed principally on fruit (Lillegraven *et al.* 1979).) Very little postcranial evidence is available but a few astragali suggest that most early didelphoids were ground dwellers (Szalay 1984). A modern dental analogue is *Marmosa* (mouse opossum) which eats a mixture of small invertebrates and fruit (Walker 1975).

Most of these marsupials became extinct at the K/T boundary, at least in the Northern Hemisphere. The multituberculates continued undiminished through the Palaeocene where they were joined by the placental carpolestids (Plesiadapiformes), the teeth of which show strong convergent morphology with those of the ptilodontoid multituberculates and a similar diet is inferred (Biknevicius 1986). Related plesiadapids show evidence of scansorial locomotion; they probably also became frugivorous but there is much disagreement on details (Szalay & Delson 1979). The only distinct specializations for eating fleshy fruit in the Palaeocene are in two picrodontid genera (also Plesiadapiformes) with teeth like those of modern fruit bats. Three or four genera or paromomyids (recently recognized as dermopterans–colugos) are considered to be scansorial or aerial forms feeding extensively on tree exudates like modern sugar gliders (Beard 1990).

Most of the remaining Palaeocene mammals that included fruit in their diets were condylarths (primitive ungulates). Many are known only from jaws and teeth but the available postcranials suggest that they were ground dwellers. The earliest and probably most specialized for frugivory were the periptychids which had massive conical ridged premolars and dominant puncture-crush mastication. Hard seeds may have formed an important part of their diet (Rensberger 1986).

(b) *Plants*

Fossil palynofloras and leaf floras from mid- to high-northern palaeolatitudes show the diversification and rise to dominance of angiosperms during the mid Cretaceous (from near 0 % at 120 Ma to 50–80 % at 66 Ma at the end of the Cretaceous) (Lidgard & Crane 1990). A broader assessment of palynofloras from 80° N to 20° S (Crane & Lidgard 1990) also shows this pattern with diversifications of magnoliid dicotyledons and monocotyledons followed rapidly by non-magnoliid dicotyledons which account for most early angiosperm diversity. Evidence from flowers shows that the Cretaceous magnoliids included Chloranthaceae, Lauraceae, and forms close to extant Magnoliaceae; with pollen evidence alone also indicating Winteraceae and possibly Myristicaceae (Friis *et al.* 1987; Crepet *et al.*, this symposium; Crane & Lidgard 1990; Crane *et al.* 1989; Drinnan *et al.* 1990; Muller 1985; Doyle *et al.* 1990). Cretaceous non-magnoliids (hamamelids and rosids) included Platanaceae and Trochodendrales (Crane 1989).

Most Cretaceous leaf floras are much in need of systematic revision (Wolfe & Upchurch 1987) but include a range of Lauralean, Magnolialean and Nymphaealean forms along with forms referable to Trochodendrales and other Hamamelidae, platanoids and primitive Rosidae (sapindophylls) (Friis *et al.* 1987; Crane 1989, Crane *et al.* 1989, Crane & Blackmore 1990; Crabtree 1987). In the later Cretaceous flacourts and euphorbs, palms and *Fagopsis* (Fagaceae) are also important (Wolfe & Upchurch 1986; Crabtree 1987). In a recent revision using foliar architecture and cuticular anatomy, Upchurch &

Dilcher (1990) recognize 70 % of a Cenomanian leaf flora to be of magnoliid form (Magnoliales, Laurales and Illiciales). None of these leaves could be assigned to an extant family with certainty.

The late Cretaceous angiosperm-dominated vegetation included shrubs, small trees and large trees (Wolfe & Upchurch 1987). In North America, trunk woods from south-eastern sites (maximum diversity five species) show little evidence of seasonality and conductivity values suggest a range through large, medium-sized and small trees. Several are represented by trunk pieces at least 1 m in diameter. A late Cretaceous wood assemblage from California (Page 1981) is more diverse (47 angiosperm species). Most again show no growth rings, but although conductivity values show that some may have been large trees, most probably represent small trees or shrubs (Wolfe & Upchurch 1987). Seventy percent of the 100 sufficiently well-preserved specimens (of 200) represented small stems, branches or roots (Page 1981). Some of these words are sufficiently distinct to suggest close relation with modern groups. These include platanoids, lauracioids and forms similar to modern Sapindales and Symplocaceae.

Direct evidence from fossil fruits and seeds in late Cretaceous and early Palaeocene floras suggests predominance of small, dry seeds and nutlets (e.g. Platanaceae, Juglandaceae, Betulaceae, Trochodendrales, Magnoliales, Cercidiphyllceae, Hamamelidaceae, Ulmaceae and Theaceae) along with some small drupes (single-seeded stone fruits; e.g. Cornaceae, Icacinaceae, Aquifoliaceae, Sabiaceae, Sapindaceae and rarer Ulmaceae (Celtidoideae), Mensipermaceae and Anacardiaceae) probably with little leathery flesh (Knobloch & Mai 1986; Mai 1987; Collinson 1990; Crane & Blackmore 1990; Friis *et al.* 1987). Angiosperms were of low diversity compared with those of the Eocene (Collinson 1990). The subordinate conifers and other seed plants included similar seed forms, for example, dry seeds of Taxodiaceae or the fleshy seeds of *Ginkgo* (Crane *et al.* 1990) (comparable in potential palatability to a drupe). Close-living relatives of leaf, pollen and wood floras discussed above do not imply the presence of other fruit types. Rare fleshy fruits are represented by an early Palaeocene lauracioid (M. E. Collinson, unpublished data) and late Palaeocene *Psidium* (Myrtaceae) and a 'fleshy fruit' (Crane *et al.* 1990). Vitaceae seeds also occur in the late Palaeocene (Mai 1987).

(c) *Interactions*

Wolfe & Upchurch (1987) interpret late Cretaceous vegetation as open canopy broad-leaved evergreen woodland which was replaced by rain forest in the early Palaeocene. Modern mammals with arboreal specializations similar to the multituberculates (e.g. most South American didelphids) are confined to dense forest. The North American Virginian opossum is, however, an exception and extremely adaptable regarding habitat. The expansion of multituberculates with angiosperms in the Cretaceous, their continuation through into the Palaeocene and the strong convergent

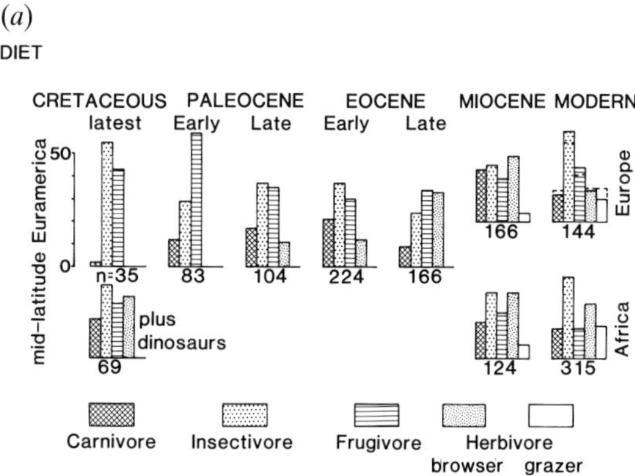

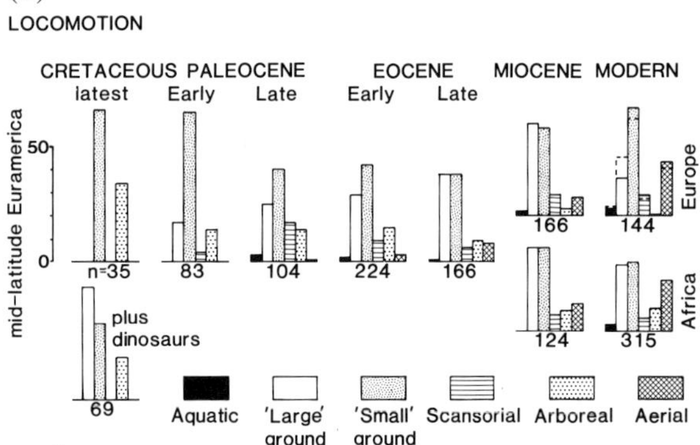

Figure 2. Ecological diversity histograms for dietary (*a*) and locomotor (*b*) adaptation, in mammal faunas from latest Cretaceous to modern, and in latest Cretaceous, the sum of mammal plus dinosaur faunas ('plus dinosaurs' histogram). Percentages of different dietary classes are shown, after the methodology of Andrews *et al.* (1979) and Collinson & Hooker (in Friis *et al.* 1987). Species of mixed classes are scored proportionally. Faunas are composite and originate from north hemisphere mid-latitudes as these are best known and easiest to interpret. The inclusion of European (temperate) and African (tropical) faunas for the Miocene and modern only shows that the pattern is independent of latitudinal control. Sources: latest Cretaceous: Lance Formation, Wyoming, U.S.A. and Scollard Formation, Alberta, Canada (Lillegraven *et al.* 1979; Weishampel *et al.* 1990); early Palaeocene: composite puercan, western U.S.A. (Savage & Russell 1983); late Palaeocene: Clarkforkian, Wyoming, U.S.A. (Rose 1981) and Cernay, France (mainly Russell *et al.* 1982); early Eocene: Wasatchian, Wyoming, U.S.A. (Schankler 1980) and European Sparnacian (mainly Russell *et al.* 1982); late Eocene: 15 English, French and Swiss sites (Hooker in Prothero (1991)); European Miocene: Pasalar, Turkey (Andrews 1990), la Grive (modified from Guérin & Mein (1971)) and Sansan (P. Andrews, unpublished data), France; African Miocene: Chamtwara, Fort Ternan, Koru and Songhor (P. Andrews, unpublished data); modern Europe (Corbet & Ovenden 1980); modern Africa: 23 low latitude sites covering a wide range of habitats (Andrews *et al.* 1979, pp. 184–185; P. Andrews, unpublished data). *n*, Number of species. Dashed lines on modern European histogram incorporates extinct mainland late pleistocene species (Kurtén 1968). This increases the similarity with tropical Africa but does not radically alter the histogram.

adaptation of carpolestids to a diet of mainly dry fruits suggests that the evolution of these mammals was linked to that of the angiosperms. Early Palaeocene GM periptychids had a similar diet. Mammalian and angiosperm fruit and seed dispersal interactions (Howe 1986; Estrada & Fleming 1986) involving small seeds carried or defecated away from the parent plant, may have commenced at this time, although much of the interaction probably resulted in seed predation rather than dispersal. Scatterhoarding usually involves larger dry fruits (Estrada & Fleming 1986; Vander Wall 1990) and thus is less likely to have been involved in the Cretaceous (see below).

Late Cretaceous and early Palaeocene floras evidently included angiosperm trees and shrubs on which were borne a variety of palatable leaves. These, in contrast to their fruit and seeds, were apparently not exploited by mammals at this time.

5. PALAEOGENE EVOLUTION OF NEW STRATEGIES

The evolution of new strategies may be considered by examining changes in both diet and locomotor adaptations. These are part of the broader study of ecological diversity (Fleming 1973; Andrews *et al.* 1979; Collinson & Hooker in Friis *et al.* (1987)). Adaptations are assessed for mammalian faunal assemblages and plotted as histograms (figure 2). Changing patterns in locomotor adaptation within frugivory and herbivore browsing are also illustrated (figure 3, see later). Faunas are presented as composite data, in an attempt to avoid bias from local communities (in particular habitats), but also to determine general trends (in contrast to our detailed studies of southern England (Collinson & Hooker in Friis *et al.* (1987); Hooker in Prothero (1991)).

(*a*) *Dietary adaptation* (*figure 2a*)

As noted above, late Cretaceous and early Palaeocene plant-eating mammals were frugivores. In fact, most plant-eating mammals developed first as frugivores, presumably because fruit is more easily processed than foliage. Mammals did not invade the herbivorous niche until the Middle Palaeocene. Evolution of large size was a prerequisite for the exploitation of leaves because longer residence time in the gut for bacterial fermentation is required to obtain sufficient nutrients from leaves. Small herbivorous rodents can exist today only because they have reduced the required gut residence time by more efficient mechanical preparation by the teeth (Janis & Fortelius 1988).

In the late Cretaceous, dinosaurs occupied the herbivorous niche and, with their addition, the late Cretaceous histogram resembles that of the Miocene or modern tropics except for the absence of grazers. These do not appear until much later, their explosive radiation being in the Miocene, coincident with a similar radiation of grassland-forming grasses (Thomasson & Voorhies 1990).

For mammals alone the late Cretaceous histogram is very different, being dominated by insectivores and frugivores with very few carnivores. After dinosaur extinction the early Palaeocene shows an increase in frugivores and a decrease in insectivores, indicating more specialization for plant-eating. Carnivores also diversify, as mammals increase in size after the extinction of the dinosaurs.

Herbivore browsers first appear in the Middle Palaeocene but they did not become significant until the late Eocene (see later). Frugivory declines first, with the appearance of herbivore browsing, and again with the appearance of grazers. The post-Miocene increase of grazers occurs at the expense of herbivore browsers.

(b) *Locomotor adaptation (figure 2b)*

In this case, even including the dinosaurs in the late Cretaceous plot does not make it similar to the modern because dinosaurs add only a strongly terrestrial (LGM) element. (We have not included aerial pterodactyls as they occur in marine settings distant from the mammal occurrences.) For mammals, only arboreal and 'small' ground classes existed in the late Cretaceous. 'Large' ground and scansorial mammals appeared in the early Palaeocene after dinosaur extinction; aquatic forms appeared in the Middle Palaeocene and aerial forms in the late Palaeocene. Aerial mammals became abundant by the late Eocene with the diversification of bats, which had appeared in the latest Palaeocene.

SGMs dominate over LGMs until the late Eocene, when LGMs increase to equivalent status. In modern temperate regions SGMs again dominate even when the extinct Pleistocene megafauna is included (dashed histograms). Arboreal forms are slightly reduced in the late Eocene and more markedly so from the Miocene onwards in temperate latitudes.

6. FRUGIVORY

(a) *Changes in locomotor adaptation within frugivory (figure 3a)*

At the Cretaceous–Tertiary boundary the first 'large' ground frugivores appear. Marsupials were essentially replaced by placentals that rapidly underwent a radiation from almost entirely insectivorous 'small' ground mammals. Many mixed frugivory with insectivory or carnivory, but others, like periptychids, were more specialized frugivores. Placentals also replaced most marsupials as SGM frugivores. LGM frugivores show a progressive reduction from early Eocene to modern, irrespective of latitude. The proportion of SGM frugivores is also reduced at low latitudes by the expansion of the arboreal primates and the aerial bats.

The late Palaeocene increase in scansorial frugivores is due to radiation of the (now extinct) primate-like Plesiadapiformes. The presence of this group reduced the proportions of LGM and SGM which resumed their previous importance with the near extinction of the Plesiadapiformes at the Palaeocene–Eocene boundary and the appearance of artiodactyls and perissodactyls.

The arboreal frugivore proportions have remained more or less stable to the present day, except where greatly reduced in modern temperate faunas.

The major category of aerial frugivores today, the fruit bats (Howe 1986; Marshall 1975; Estrada & Fleming 1986), do not appear until the Miocene when essentially modern forms are found in southern France and tropical Africa (Sigé & Aguilar 1987). Fossil fruits and seeds, the close living relatives of which are an important part of fruit bat diet (Marshall 1975), are known from the Kenyan Miocene (Chesters 1957; M. E. Collinson, personal observations). These include Annonaceae, Lauraceae, Tiliaceae, (*Grewia*), Sterculiaceae, Cucurbitaceae, Combretaceae, Euphorbiaceae, Rhamnaceae, Vitaceae, Sapindaceae, Burseraceae, Anacardiaceae, Meliaceae, Ehretiaceae, Rubiaceae and Arecaceae (*Phoenix*). Many of these are also known from much earlier fossils (see elsewhere herein) so do not reflect coevolution with bats.

(b) *'Large' ground frugivory*

The presence of quite high numbers of 'large' ground mammalian frugivores in the Palaeogene may

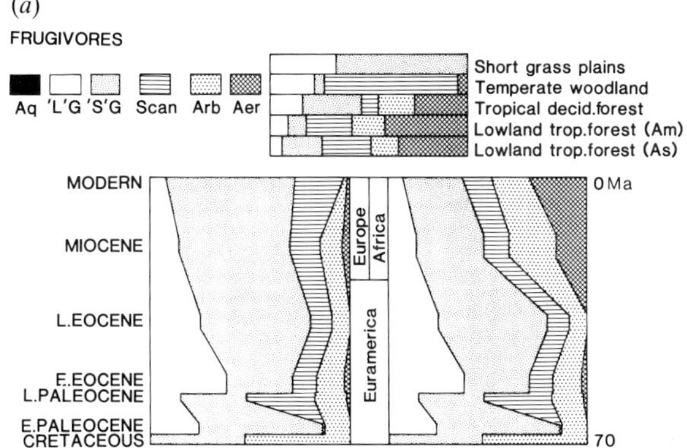

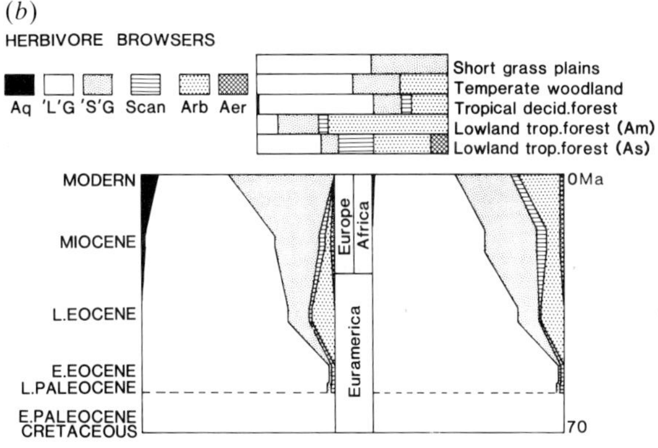

Figure 3. Proportions of locomotor classes among (*a*) frugivores and (*b*) herbivore browsers through time (faunas in figure 2). Plots for five modern habitat types (horizontal bar diagrams at top of each figure) show that changing vegetation is not the major control over past pattern, although tropical deciduous forest fits reasonably well with African Miocene. Faunas from short grass plains and tropical deciduous forest from Andrews (unpublished data); temperate woodland and American (Am) lowland tropical forest from Fleming (1973); S.E. Asian (As) lowland tropical forest from Davies (1962). Ma = millions of years before present. Note that the two diagrams at left and right are the same up to the late Eocene, whereas above that level separate versions are shown for Europe and Africa.

seem surprising in view of their rarity today. Modern examples are peccaries, agoutis and chevrotains. The proportion of 'large' ground frugivores falls from the late Palaeocene to the present day, regardless of latitude or of the increase in arboreal and aerial frugivores in the tropics.

Conclusive evidence for the terrestrial nature of Palaeogene frugivores is provided by the skeleton of the artiodactyl *Aumelasia* from Messel which has bunodont teeth like those of *Acotherulum* (figure 4c). The tooth gross morphology, gross wear and microwear of the latter all compare closely with fruit-eating monkeys. Cusps are rounded, there is little cresting and puncture-crush wear is much in evidence on the cusp tips. The gut contents of *Aumelasia* include fruits, seeds and sand grains which could only have been ingested when picking up fallen fruit (Franzen & Michaelis 1988).

Possible reasons for the subsequent demise of this strategy are as follows.

1. Reduction in fallen fruit abundance which may be due to a reduction in climatic equability resulting in alteration of patterns of fruit production (cf. Terborgh in Estrada & Fleming (1986)). Insufficient fallen fruit in the neotropics during the season of scarcity is a population-regulating factor for terrestrial frugivores today (Smythe 1986).

2. Competition with arboreal frugivores taking fruit at source. Multituberculates, in taking mainly dry fruits and seeds, may not have caused the same degree of competition as later primates.

3. Competition with rodents (especially outside the tropics) able to open all but the hardest of dry nuts and seeds, and which, because of their scatterhoarding (Estrada & Fleming 1986; Vander Wall 1990) might have been selected for as plant dispersers.

(c) *Primate frugivory*

Our frugivorous category incorporates any fruit or seed eating. A combination of gross dental morphology and microwear can be used to provide some indications of types of frugivory i.e. soft fleshy versus hard dry. As noted earlier few mammals in the late Cretaceous and Palaeocene had distinct adaptations for eating fleshy fruit.

True primates appeared just before the Palaeocene–Eocene boundary in North America, Europe, Asia and North Africa. Today they are amongst the most important arboreal frugivores (Fleming *et al.* 1987; Howe 1986; MacKinnon 1978; Chivers 1980; Estrada & Fleming 1986). Initially primates were insectivorous, then combined insectivory and frugivory. In the late Early and Middle Eocene of Europe many primates (especially in the family Adapidae) were frugivorous. Their molars are typified by low blunt cusps showing much wear at the tips; their wear facets bear only very fine striations and very little pitting (observed by light microscopy). This suggests a fleshy fruit diet. It coincides with the widespread extent of tropical to paratropical forests reaching to 50°–60° North and South (Wolfe 1985; and see Collinson (1990) for review) represented in the fossil record by a diversity of fruits and seeds (Collinson 1983, 1990, in Franzen &

Michaelis (1988), in Crane & Blackmore (1989); Mai & Walther 1985; Manchester in Knobloch & Kvaček (1990)) whose close living relatives include fleshy fruits important in the diets of primates today (Terborgh in Estrada & Fleming (1986); Chivers 1980; Gautier-Hion *et al.* 1985; Leighton & Leighton 1983). These include Annonaceae, Moraceae (especially *Ficus*), Vitaceae, Sapindaceae, Sapotaceae, Anacardiaceae, Euphorbiaceae, Burseraceae. The significance of several of these families in ancient vegetation of this time is supported by leaf and pollen fossils (Muller 1985; Wilde 1989; Manchester in Knobloch & Kvaček (1990)). Confirmation of the former existence of softer tissues in fruits is available, for example for Menispermaceae, Vitaceae, Lauraceae and Cornaceae at Messel (Collinson in Franzen & Michaelis (1988)), and in some permineralized fossils (e.g. Lauraceae in the London Clay flora (Collinson 1983)).

Many factors influence fruit and seed selection, and dispersal versus predation, in frugivory. These include fruit colour and composition of pulp, particularly with respect to lipids, proteins and sugars (Howe 1986; Estrada & Fleming 1986). These factors are unknown for fossils. Physical features can be observed. These include (rarely) pericarp texture and thickness and in almost every case the thickness and strength of the layer which would have been the major protection for the embryo (testa, endocarp, etc.). In most cases the ancient examples are very like their close living relatives and they offer no evidence for modification in response to frugivory.

The family Lauraceae is extremely abundant and diverse in these tropical or paratropical Eocene forests (Collinson 1983, in Franzen & Michaelis (1988); Wilde 1989; Mai & Walther 1985). Modern Lauraceae exemplify one extreme fruiting strategy, having large fruit, high in protein and lipid content (termed high quality, high investment fruit). In consequence fruits of many modern Lauraceae are dispersed by birds, although they can form a very subordinate element in some primate diets, e.g. howler and capuchin monkeys (Howe 1986; Fleming *et al.* 1987; Estrada & Fleming 1986, Herrera (in Estrada & Fleming 1986) quoted *Laurus* species from distinct modern habitats where fruits were identical (colour, form, composition) but were dispersed by quite unrelated birds, as evidence against any close coevolutionary relation between the fruit and bird dispersal agents. The fossil record of birds is fragmentary but there is no evidence to suggest that a wide range of frugivorous birds existed in the European early Eocene (Harrison 1979). The Eocene fruits are mostly only medium-sized and it is possible that they formed part of the arboreal mammalian frugivore diet.

Cornaceae are another group with mainly bird dispersers today although some mammals also take the fruit (Eyde 1988). The family (here taken to include Mastixiaceae) has an extensive fossil history with fruits recorded from the late Cretaceous onwards (Eyde 1988). Late Cretaceous and Palaeocene forms were restricted to mastixioids probably with scant leathery flesh. These were small in the late Cretaceous (Knobloch & Mai 1986) but one early Palaeocene example

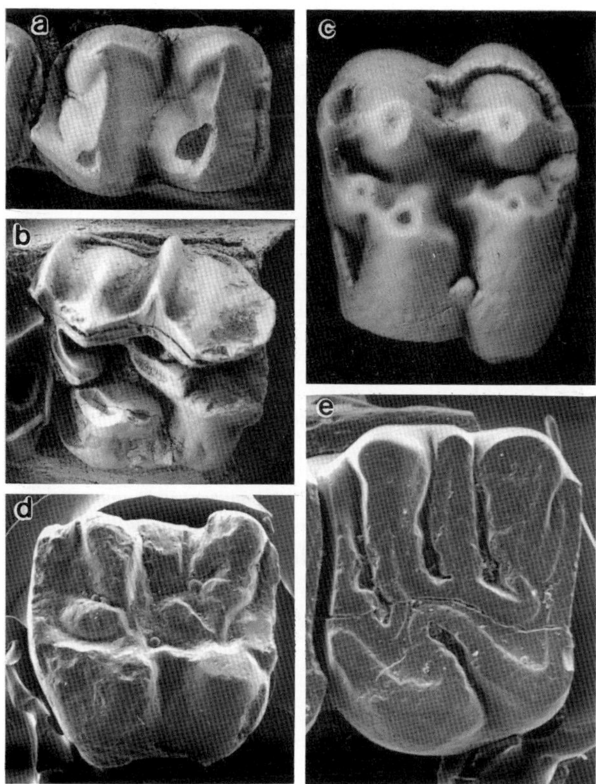

Figure 4. Upper (b–e) and lower (a) pre-ultimate molars of fossil mammals from the Eocene of southern England. (a–c) LM, coated with ammonium chloride; (d–e) SEM. Numbers refer to specimens in the Palaeontology Department, Natural History Museum (BM(NH)). (a) *Hyrachyus stehlini*, a bilophodont herbivore browser (perissodactyl), M50194, × 2.5. (b) *Plagiolophus annectens*, a selenodont semihypsodont herbivore browser (perissodactyl), 29715, × 2.3. (c) *Acotherulum quercyi*, a frugivore (artiodactyl), M29470, × 4.8. (d) *Sciuroides rissonei*, a frugivore (soft) (rodent), M37368, × 12. (e) *Thalerimys headonensis*, a semihypsodont herbivore browser (rodent), 30159a, × 18.

(Puercan, Colorado) is large (endocarp more than 20 mm long) (M. E. Collinson, unpublished results). Eocene forms included those with thicker soft tissues (Eyde 1988). Eocene fruits are typified by higher locule number than modern examples whereas Miocene forms often have thicker, strongly ornamented endocarps (Eyde 1988; Knobloch & Mai 1986). Rodent-gnawed examples occur in the Miocene (Boucot 1990, p. 263). A thorough survey of fruit trends in this family, so well represented in the fossil record, would be worthwhile. They too might have been food for early mammalian frugivores.

(d) *Rodent frugivory*

Rodents are another important group of frugivorous mammals today, tending to be specialized for feeding on hard dry fruit types (Estrada & Fleming 1986; Vander Wall 1990).

Rodents first appeared in the late Palaeocene: about 1 Ma earlier in North America than in Europe but in both cases coincident with drastic reduction in multituberculates (Krause 1986). However, the locomotor adaptations of the rodents appear to have been considerably less specialized for life in the trees than

were the multituberculates (Szalay 1985). Generally they lack specific arboreal modifications and we have scored them as SGM/scansorial.

Rodents underwent a relatively modest radiation during the early and Middle Eocene, although they were abundant as individuals in nearly all faunas. They underwent a major radiation in the late Eocene to become a major component of faunas, as they are today. Many remained SGM/scansorial although some later became LGM, others arboreal and some partly aerial.

The early rodents would have been able to exploit the variety of dry seeds and fruits (see above) which had been exploited by multituberculates. Krause (1986) argued strongly for competitive exclusion of multituberculates by rodents. One major adaptive advantage possessed by rodents was evergrowing incisors, not possessed by the gnawing taeniolabidoid multituberculates. Angiosperm families that produce large, rodent-dispersed nuts today (Betulaceae, Fagaceae, Juglandaceae) were producing mainly smaller (often winged) nutlets during the late Palaeocene and earliest Eocene. The first production of the larger nuts in these families occurs during the latest Palaeocene or the early and Middle Eocene and they were not abundant until the late Eocene (Crane & Blackmore 1989). An earlier rodent expansion may have been contained by the abundance of soft fruit more suited to exploitation by primates in tropical forests during the early and Middle Eocene but as these forests became more restricted the deciduous nut-bearing families diversified (Crane & Blackmore 1989; Wolfe in Prothero (1991); Collinson in Prothero (1991)).

It is possible that these early rodents hoarded nuts, as do modern rodents. The earliest evidence of nut hoarding by rodents appears to be late Miocene heteromyid burrows containing *Celtis* endocarps (Boucot 1990, p. 57). *Celtis* endocarps are known in the Palaeocene and are abundant from the Eocene onwards (Manchester in Crane & Blackmore (1989)).

Complex interactions between fruit and scatter-hoarders are discussed by several authors in Estrada & Fleming (1986). For example Stapanian describes the relative value of *Juglans*, *Carya* and *Quercus*; the two former being more digestible and having a higher lipid content but being thick walled, whereas the latter yields more energy per unit time (even though twice the mass is needed to compensate for lower nutrient content) because the wall is thin. Relative sizes of nuts influence distance carried; this and timing of germination influence recovery and predation. Platt & Hermann show that the scatterhoarding of these species is essential for their persistence as codominants in a subtropical forest in northern Florida. Papers by Janzen and Hallwachs discuss the role of rodents in removing seeds from dung of megaherbivores. Here the size of seed and its visibility in dung are important. Miocene rodent-gnawed mastixioid Cornaceae (Boucot 1990, p. 263) would have possessed pulp so may have been secondarily predated by rodents.

Some aspects of significance to these rodent dispersal factors such as size of nuts and wall thickness can be assessed from the fossil record; especially that of

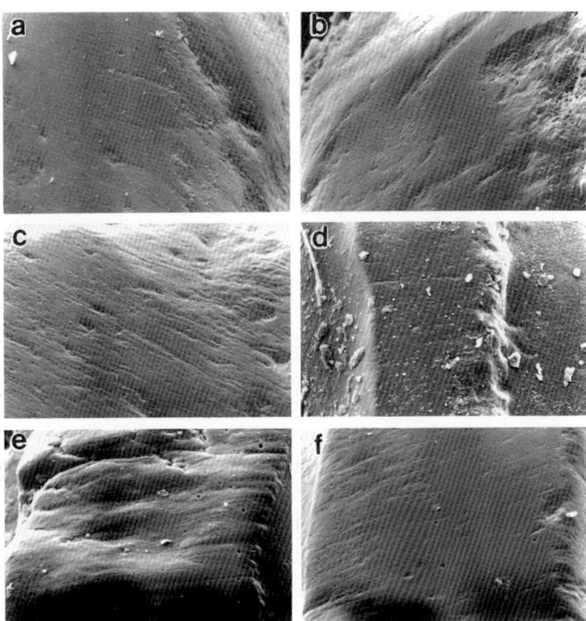

Figure 5. SEMS of microwear of fossil and modern mammal upper pre-ultimate molars (casts); buccal phase. Numbers refer to specimens in the Natural History Museum (BM(NH)); ZD indicates modern in Zoology Department, M or no prefix indicates fossil in Palaeontology Department. (a) *Ratufa affinis*, ZD9.4.1.218; (b) *Sciuroides rissonei*, (figure 4d); (c) *Treposciurus mutabilis*, M51685; 9d) *Capreolus capreolus* (modern); (e) *Alcelaphus buselaphus* (modern); (f) *Plagiolophus annectens*, M52683 (cf. figure 4b). (a–c) ×200; (d–f) ×60.

Juglandaceae. Detailed studies would be worthwhile to help elucidate the role of early rodents as seed dispersers.

(e) *Evidence of diet in southern England*

Some of the early and Middle Eocene rodents seem to have exploited the soft fruit availability at this time. Indeed, in southern English faunas (Collinson & Hooker in Friis *et al.* (1987)) and also in Europe we see a trend from softer to harder diet in rodents during the Eocene. A modern dental analogue for pseudosciurids (figure 4d) and manitshine paramyids in the middle Eocene is the giant squirrel (*Ratufa*). These eat both soft and hard fruit and a few flowers (MacKinnon 1978). The teeth have square outlines (which gives an increased crushing area) with low blunt cusps.

We have examined the microwear of one Middle and one late Eocene Pseudosciurid genus, shown here (figure 5) with *Ratufa* for comparison. This seems to indicate that *Sciuroides* (figure 5b) with few scratches on overall very smooth facets was eating a softer food than *Treposciurus* (figure 5c) with many, stronger, parallel striations and more pits. The modern *Ratufa* (figure 5a) is similar to *Sciuroides*.

This dietary modification coincides with reduction in individual abundance and species diversity of pseudosciurids at the Middle–late Eocene boundary; followed by a modest radiation of new types in the late Eocene with coarser diet including some fruit and some leaves. The frugivorous pseudosciurids were replaced by theridomyids (see herbivore browsing) and glirids (Collinson & Hooker in Friis *et al.* (1987)).

The small glirids (dormice) have low-crowned molars characterized by low ridges with enhanced grinding capacity suggesting a frugivorous diet different from that of the pseudosciurids. The plant items in the diet of modern glirids comprise a range of hard and soft fruits (Walker 1975). We have evidence for at least part of the diet for one species, as seeds of *Stratiotes* (with appropriate-sized gnaw marks) occur at two distinct late Eocene horizons (Collinson 1990); at one of these, in association with a glirid tooth.

7. HERBIVORE BROWSING

(a) *Changes in locomotor adaptation within herbivore browsing (figure 3b)*

The earliest herbivore browsers appeared in the Middle Palaeocene; almost all were LGMs. Pantodonts dominated with dinocerates and tillodonts in the later Palaeocene. They appear to have developed herbivory directly from insectory by increase in size and little initial tooth modification other than increased cresting.

Early Eocene herbivore browsers were also mainly LGMs but Palaeocene forms were replaced by new groups which evolved from LGM frugivores; especially perissodactyls and, as the Eocene proceeded, also artiodactyls. This contrasts with the picture just presented for frugivores where major changes in locomotor category occurred at this time.

LGM herbivore browsers show a decline from early Eocene to present in Europe and Africa; but not as marked as the LGM frugivore decline. This is balanced by an expansion of SGM and arboreal herbivore browsers in the Eocene. Most arboreal forms in the late Eocene were primates whose diet had shifted from frugivory (Szalay & Delson 1979). After this the arboreals declined and the SGMs continued to expand in Europe but both continued expanding in Africa.

(b) *Large ground herbivore browsing*
(i) *Bilophodonty*
The simplest dental modifications for leaf eating were developed in Eocene herbivores such as the rhinocerotoid *Hyrachyus* and occur today in tapirs and tree kangaroos. These teeth (figure 4a) had straight transverse crests (usually two per tooth) and are termed bilophodont. These performed mainly a shearing action slicing up leaves into quite large pieces. Bulk food was processed rapidly and inefficiently, a method typically used by perissodactyls to exploit mainly cell contents (Janis 1989). The modern tapir diet is restricted almost entirely to leaves of forest trees (Terwilliger 1978).

(ii) *Selenodonty*
A major radiation of LGM herbivore browsers occurred in the late Eocene (Collinson & Hooker in Friis *et al.* (1987)). Probable fossil grasses go back to the Eocene (Thomasson & Voorhies 1990). However, the earliest fossils securely attributed to the grasses are late Oligocene, and rapid diversification and spread of grasslands was not until the Middle Miocene. Fossil soils similar to those of modern grasslands have also

been found in the Oligocene (Retallack 1986). In the light of these reports, and the fact that horse evolution is intimately connected with grazing adaptation, it is relevant to look at *Plagiolophus* (a palaeothere, figure 4b), which is both an equoid and the highest crowned (semi-hypsodont) herbivore in the European late Eocene, to see if its microwear suggests it was a browser or a grazer. Hypsodonty is generally associated with a more abrasive diet.

We have compared the microwear of *Plagiolophus* with *Capreolus* (roe deer), a moderately high-crowned browser and *Alcelaphus* (hartebeest) a grazer. *Alcelaphus* (figure 5d) shows a typical coarse-scale irregular pattern with fine-scale strong striations. *Capreolus* (figure 5e) shows a flat surface with no coarse pattern and a fine-scale pattern of fine striations. *Plagiolophus* (figure 5f) is very similar to *Capreolus* and thus a browser.

(c) 'Small' ground herbivore browsing

SGM herbivore browsers expanded from the late Eocene to the present day. Many of these are rodents such as theridomyids (e.g. *Thalerimys*, figure 4e). These are semi-hypsodont which implies the need to process a greater quantity of food than low-crowned pseudosciurids.

We can conclude that most theridomyids were eating a high proportion of leaves. A modern dental analogue, *Thryonomys* (the cane rat) (Collinson & Hooker in Friis *et al.* (1987), eats the roots and shoots of coarse grasses and shrubs as well as nuts, fallen fruit and bark (Walker 1975).

8. DISCUSSION AND CONCLUSIONS

In the case of frugivory two processes have operated during the Cainozoic. There has been a shift in locomotor adaptation (from the ground into the trees and the air) and a shift in diet (from insectivory or carnivory to frugivory in nearly all the locomotor classes). This implies pressure to change locomotor adaptation to obtain fruit at source and to change dietary adaptation to exploit a new food resource.

The situation is different for browsing herbivores. The earliest herbivores were ground dwelling and achieved their dietary specializations mainly through evolution from already large, ground-dwelling frugivores (or, in the Palaeocene, by size increase from small insectivorous ancestors). Large size limited them to the ground. Most browsing herbivores in other locomotor niches changed their diet from frugivory without changing their locomotor adaptation. Notable exceptions today are the tree kangaroo and the tree hyrax. The dietary shifts probably resulted from reduction in fruit availability, especially in the late Eocene.

A period of nearly 30 Ma existed between the dominance of angiosperms and their exploitation by mammalian herbivores. Even after dinosaur extinction several million years intervened without any evidence for dietary interaction between angiosperms and vertebrate herbivores. Expansion of this interaction did not occur until the late Eocene.

The delay in developing herbivory may have

resulted from the early Palaeocene equable climate which turned late Cretaceous open canopy woodland into rainforest (Wolfe & Upchurch 1987; Wolfe 1990) with an ample supply of fruit and dense vegetation which might have discouraged evolution of large size in mammals. We are unsure how much influence the removal of dinosaur herbivory had on plant communities following the K/T boundary (Friis *et al.* 1987) but the strong temporal link between the various later radiations of herbivore specializations and climate-induced vegetational changes seem inescapable (herein; Collinson & Hooker in Friis *et al.* (1987); Janis 1989). Definite examples of coevolution seem elusive but we can suggest the timing of certain events that may represent the beginnings or major development of interactions which some authors consider examples of coevolution. Opinions vary greatly on coevolution in fruit and seed dispersal (Estrada & Fleming 1986). One extreme is represented by Janzen's and Hallwach's claims (p. 252 and p. 285 respectively in Estrada & Fleming (1986)) for coevolution between large legume seeds like *Hymenaea* (Guanipol) and *Enterolobium* (Guanacaste) with extinct megaherbivores in South America. The other is favoured by many workers on avian frugivory and is specifically supported by documentary evidence by Herrera (p. 14 in Estrada & Fleming (1986)) who argues: 'insofar as plant adaptations are concerned, there exists little or no basis for assuming generally gradualistic change and environmental tracking in evolutionary formulations of plant-vertebrate seed disperser interactions'.

Possible examples of diffuse coevolution are: the late Cretaceous and Palaeocene expansion of arboreal dry fruit- and seed-eating multituberculates and angiosperm-dominated vegetation; the late Palaeocene appearance and late Eocene radiation of large nuts and rodents; the early to Middle Eocene expansion of fleshy fruits and the evolution of frugivorous arboreal primates.

Three features characterize the late Eocene. There is a change in frugivore diet from soft to hard fruit (to some extent a reversal of early Eocene trends); dietary shifts in nearly every locomotor class from frugivory to herbivore browsing and the existing herbivore browsers show evidence of increasing coarseness of diet. In addition, 'large' ground mammals reach equivalent abundance to small ground mammals and there is a slight reduction in arboreal forms. These changes are all consistent with patterns of vegetational and climatic change during the late Eocene in which previously widespread megathermal, tropical to paratropical, broad-leaved evergreen forests became increasingly restricted to low latitudes; being replaced in mid-high latitudes by mesothermal to microthermal, mixed evergreen and deciduous forests (Wolfe 1985, 1987; Collinson 1990; Wolfe in Prothero (1991); Collinson in Prothero (1991)).

We thank Peter Andrews, Mikael Fortelius and Jack Wolfe for helpful discussion. Peter Andrews also made available unpublished information and critically read the text. This work was undertaken by M. E. C. while in receipt of a Royal Society 1983 University Research Fellowship, which is gratefully acknowledged.

REFERENCES

Andrews, P. 1990 Palaeoecology of the Miocene fauna from Paşlar, Turkey. *J. hum. Evol.* **19**, 569–582.

Andrews, P., Lord, J. M. & Nesbit Evans, E. M. 1979 Patterns of ecological diversity in fossil and modern mammalian faunas. *Biol. J. Linn. Soc.* **11**, 177–205.

Beard, K. C. 1990 Gliding behaviour and palaeoecology of the alleged primate family Paromomyidae (Mammalia, Dermoptera). *Nature, Lond.* **345**, 340–341.

Biknevicius, A. R. 1986 Dental function and diet in the Carpolestidae (Primates, Plesiadapiformes). *Am. J. Phys. Anthrop.* **71**, 157–171.

Boucot, A. J. 1990 *Evolutionary paleobiology of behaviour and coevolution.* Amsterdam: Elsevier.

Chesters, K. I. M. 1957 The Miocene flora of Rusinga Island, Lake Victoria, Kenya. *Palaeontographica* B **101**, 30–71.

Chivers, D. J. (ed.) 1980 *Malayan forest primates.* New York: Plenum.

Collinson, M. E. 1983 *Fossil plants of the London Clay.* Palaeontological Association Field Guides to Fossils No. 1. London: The Palaeontological Association.

Collinson, M. E. 1990 Plant evolution and ecology during the early Cainozoic diversification. *Adv. Bot. Res.* **17**, 1–98.

Corbet, G. & Ovenden, D. 1980 *The mammals of Britain and Europe.* London: Collins.

Crabtree, D. R. 1987 Angiosperms of the northern Rocky Mountains: Albian to Campanian (Cretaceous) megafossil floras. *Ann. Missouri Bot. Gard.* **74**, 707–747.

Crane, P. R. 1989 Palaeobotanical evidence on the radiation of the non-magnoliid dicotyledons. *Pl. Syst. Evol.* **162**, 165–191.

Crane, P. R. & Blackmore, S. 1989 *Evolution, systematics and fossil history of the Hamamelidae,* vol. 1 (*Introduction and 'lower' Hamamelidae*); vol. 2 (*'Higher' Hamamelidae*). Systematics Association Special Volumes 40A & 40B. Oxford: Clarendon Press.

Crane, P. R., Friis, E. M. & Pedersen, K. R. 1989 Reproductive structure and function in Cretaceous Chloranthaceae. *Pl. Syst. Evol.* **165**, 211–226.

Crane, P. R. & Lidgard, S. 1990 Angiosperm radiation and patterns of Cretaceous palynological diversity. In *Major evolutionary radiations.* (ed. P. D. Taylor & G. P. Larwood), pp. 377–407 (Systematics Association Special Volume no. 42). Oxford: Clarendon Press.

Crane, P. R., Manchester, S. M. & Dilcher, D. L. 1990 A preliminary survey of fossil leaves and well-preserved reproductive structures from the Sentinel Butte Formation (Paleocene) near Almont, North Dakota. *Fieldiana Geology* n.s. **20**, 1–63.

Davies, D. D. 1962 Mammals of the lowland rain-forest of North Borneo. *Bull. natn. Mus. St. Sing.* **31**, 1–129.

Doyle, J. A., Hotton, C. L. & Ward, J. V. 1990 Early Cretaceous tetrads, zonasulcate pollen and Winteraceae. II. Cladistic analysis and implications. *Am. J. Bot.* **77**, 1558–1568.

Drinnan, A. N., Crane, P. R., Friis, E. M. & Pedersen, K. R. 1990 Lauraceous flowers from the Potomac Group (Mid-Cretaceous) of Eastern North America. *Bot. Gaz.* **151**, 370–384.

Estrada, E. & Fleming, T. H. 1986 *Frugivores and seed dispersal.* Dordrecht: W. Junk.

Eyde, R. H. 1988 Comprehending Cornus: puzzles and progress in the systematics of the dogwoods. *Bot. Rev.* **54**, 233–351.

Fleming, T. H. 1973 Numbers of mammal species in north and central American forest communities. *Ecology* **54**, 555–563.

Fleming, T. H., Breitwisch, R. & Whitesides, G. H. 1987 Patterns of tropical frugivore diversity. *A. Rev. Ecol. Syst.* **18**, 91–109.

Franzen, J. L. & Michaelis, W. (eds) 1988 Eocene Lake Messel. *Cour. Forsch.-Inst. Senckenberg* **107**, 1–452.

Friis, E. M., Chaloner, W. G. & Crane, P. R. 1987 *The origins of angiosperms and their biological consequences.* Cambridge University Press.

Gautier-Hion, A. *et al.* 1985 Fruit characters as a basis of fruit choice and seed dispersal in a tropical forest vertebrate community. *Oecologia, Berl.* **65**, 324–337.

Gingerich, P. D. (ed.) 1980 Early Cenozoic paleontology and stratigraphy of the Bighorn Basin, Wyoming. *Pap. Paleont. Mus. Paleont. Univ. Mich.* **24**, 1–146.

Guérin, C. & Mein, P. 1971 Les principaux gisements de mammifères miocènes et pliocènes du domaine Rhodanien. *Docum. Lab. Géol. Lyon. H.S.* 131–170.

Harrison, C. J. O. 1979 Small non-passerine birds of the Lower Tertiary as exploiters of ecological niches now occupied by passerines. *Nature, Lond.* **281**, 562–563.

Harrison, J. L. 1962 The distribution of feeding habits among animals in a tropical rain forest. *J. Anim. Ecol.* **31**, 53–63.

Janis, C. M. 1989 A climatic explanation for patterns of evolutionary diversity in ungulate mammals. *Palaeontology* **32**, 463–481.

Janis, C. M. & Fortelius, M. 1988 On the means whereby mammals achieve increased functional durability of their dentitions, with special reference to limiting factors. *Biol. Rev.* **63**, 197–230.

Jenkins, F. A. & Krause, D. W. 1983 Adaptations for climbing in North American multituberculates (Mammalia). *Science, Wash.* **220**, 712–715.

Kerle, J. A. 1984 The behaviour of Burramys parvus Broom (Marsupialia) in captivity. *Mammalia* **48**, 317–325.

Knobloch, E. & Kvaček, Z. 1990 *Proceedings of the Symposium Paleofloristic and paleoclimatic changes in the Cretaceous and Tertiary.* Prague: Geological Survey Publisher.

Knobloch, E. & Mai, D. H. 1986 *Monographie der früchte und Samen in der Kreide von Mitteleuropa.* Edice Rozpravy Ustredniho ustavu geologickeho 47, Geological Survey Prague.

Krause, D. W. 1982 Jaw movement, dental function and diet in the Paleocene multituberculate Ptilodous. *Paleobiology* **8**, 265–281.

Krause, D. W. 1986 Competitive exclusion and taxonomic displacement in the fossil record: the case of rodents and multituberculates in North America. *Univ. Wyoming. Contrib. Geol. Spec. Pap.* **3**, 95–117.

Kurtén, B. 1968 *Pleistocene mammals of Europe.* London: The World Naturalist, Weidafeld & Nicholson.

Leighton, M. & Leighton, D. R. 1983 Vertebrate responses to fruiting seasonality within a Bornean rain forest. In *Tropical rain forest: ecology and management* (ed. S. L. Sutton, T. C. Whitmore & A. C. Chadwick), pp. 181–196. Oxford: Blackwell Scientific Publications.

Lidgard, S. & Crane, P. R. 1990 Angiosperm diversification and Cretaceous floristic trends: a comparison of palynofloras and leaf macrofloras. *Paleobiology* **16**, 77–93.

Lillegraven, J. A., Kielan-Jaworowska, Z. & Clemens, W. A. 1979 *Mesozoic mammals, the first two-thirds of mammalian history.* Berkeley, California: University of California Press.

MacKinnon, K. S. 1978 Stratification and feeding differences among Malayan squirrels. *Malay. Nat. J.* **30**, 593–608.

Mai, D. H. 1987 Neue Früchte und Samen aus paläozanen Ablagerungen Mitteleuropas. *Feddes Repert.* **98**, 197–229.

Mai, D. H. & Walther, H. 1985 Die obereozänen Floren des Weisselster-Beckens und seiner Randgebiete. *Abh. St. Mus. Miner. Geol. Dresd.* **33**, 5–176.

Marshall, A. G. 1985 Old world phytophagous bats (Megachiroptera) and their food plants: a survey. *Zool. J. Linn. Soc.* **83**, 351–369.

Muller, J. 1985 Significance of fossil pollen for angiosperm history. *Ann. Missouri Bot. Gard.* **71**, 419–443.

Murray, D. R. 1986 Seed dispersal. Sydney: Academic Press.

Page, V. M. 1981 Dicotyledonous wood from the Upper Cretaceous of central California. III. Conclusions. *J. Arn. Arb.* **62**, 437–455.

Prothero, D. R. 1991 *Eocene-Oligocene climatic and biotic evolution.* Princeton University Press. (In the press.)

Rensberger J. M. 1978 Scanning electron microscopy of wear and occlusal events in some small herbivores. In *Development, function and evolution of teeth* (ed. P. M. Butler & K. A. Joysey), pp. 415–438. London: Academic Press.

Rensberger, J. M. 1986 Early chewing mechanisms in mammalian herbivores. *Paleobiology* **12**, 474–494.

Retallack, G. J. 1986 The fossil record of soils. In *Paleosols: their recognition and interpretation.* (ed. V. P. Wright), pp. 1–57. Oxford: Blackwell Scientific Publications.

Rose, K. D. 1981 The Clarkforkian land-mammal age and mammalian faunal composition across the Palaeocene–Eocene boundary. *Pap. Paleont. Mus. Paleont. Univ. Mich.* **26**, 1–197.

Savage, D. E. & Russell, D. E. 1983 *Mammalian paleofaunas of the world.* Reading, Massachusetts: Addison–Wesley.

Schaal, S. & Ziegler, W. (eds) 1988 *Messel – ein Schaufenster in die Geschichte der Erde und des Lebens.* Frankfurt am Main: Kramer.

Sigé, B. & Aguilar, J.-P. 1987 L'extension stratigraphique des megachiroptères dans le Miocène d'Europe méridionale. *C.r. Acad. Sci., Paris* **304**, 11, 9, 469–474.

Smythe, N. 1986 Competition and resource partitioning in the guild of neotropical terrestrial frugivorous mammals. *A. Rev. Ecol. Syst.* **17**, 169–188.

Szalay, F. S. 1984 Arboreality: Is it homologous in metatherian and eutherian mammals? *Evol. Biol.* **18**, 215–258.

Szalay, F. S. 1985 Rodent and lagomorph morphotype adaptations, origins, and relationships: some postcranial attributes analyzed. In *Evolutionary relationships among rodents – a multidisciplinary analysis.* (ed. W. P. Lucket & J.-L. Hertenberger), pp. 83–132. New York: Plenum Press.

Szalay, F. S. & Delson, E. 1979 *Evolutionary history of the primates.* New York: Academic Press.

Teaford, M. F. 1988 A review of dental microwear and diet in modern mammals. *Scanning Microsc.* **2**, 1149–1166.

Terwilliger, V. J. 1978 Natural history of Baird's Tapir on Barro Colorado Island, Panama Canal zone. *Biotropica* **10**, 211–220.

Thomasson, J. R. & Voorhies, M. R. 1990 Grasslands and Grazers. In *Palaeobiology a synthesis* (ed. D. E. G. Briggs & P. R. Crowther), pp. 84–87. Oxford: Blackwell Scientific Publications.

Upchurch, G. R. & Dilcher, D. L. 1990 Cenomanian angiosperm leaf megafossils, Dakota Formation, Rose Creek locality, Jefferson County, Southeastern Nebraska. *U.S. Geol. Surv. Bull.* **1915**, 1–52.

Vander Wall, S. B. 1990 *Food hoarding in animals.* University of Chicago Press.

Walker, E. P. 1975 *Mammals of the world,* 3rd edn. Baltimore: The Johns Hopkins University Press.

Weishampel, D. B., Dodson, P. & Osmolska, H. (eds) 1990 *The Dinosauria.* Berkeley: University of California Press.

Wilde, V. 1989 Untersuchungen zur Systematik der Blattreste aus dem Mitteleozän der Grube Messel bei Darmstadt (Hessen, Bundesrepublik Deutschland). *Cour. Forsch.-Inst. Senckenberg* **115**, 1–214.

Wolfe, J. A. 1985 Distribution of major vegetational types during the Tertiary. *Am. Geophys. Union Geophys. Mon.* **32**, 357–375.

Wolfe, J. A. 1987 Late Cretaceous-Cenozoic history of deciduousness and the terminal Cretaceous event. *Paleobiology* **13**, 215–226.

Wolfe, J. A. 1990 Palaeobotanical evidence for a marked temperature increase following the Cretaceous/Tertiary boundary. *Nature, Lond.* **343**, 153–156.

Wolfe, J. A. & Upchurch, G. R. 1986 Vegetation, climatic and floral changes at the Cretaceous–Tertiary boundary. *Nature, Lond.* **324**, 148–151.

Wolfe, J. A. & Upchurch, G. R. 1987 North American nonmarine climates and vegetation during the late Cretaceous. *Palaeogeogr. Paleoclimatol. Palaeoecol.* **61**, 33–77.

Discussion

P. W. SKELTON (*Department of Earth Sciences, The Open University, Milton Keynes, U.K.*). Because of the apparent absence of mammalian browsers in the early Palaeocene, what does Dr Collinson suppose were the bulk feeders upon vegetation at this time? Or was the world covered in mulch?

M. E. COLLINSON. Today, the most important bulk feeders on vegetation are insects. This is especially the case in tropical rainforests (Wint 1983). On the evidence from physiognomy of North American leaf floras, Palaeocene mid-latitude vegetation is inferred to have been closed-canopy forest (Wolfe & Upchurch 1987). Evidence from fossil leaf damage suggests that insects were active herbivores well before the Palaeocene (Chaloner *et al.* this symposium). Assuming that leaf litter was as actively processed then as today, it is unlikely that the absence of mammalian herbivores in the early Palaeocene (and their negligible numbers during the rest of the Palaeocene) would have meant a Palaeocene world covered in mulch. Much of the continental Palaeocene sequence of the western U.S.A. does contain extensive coals. These coals are not interpreted as reworked forest floor litter, but as former mire deposits resulting from a post-Cretaceous increase in precipitation (Wolfe & Upchurch 1987).

Reference

Wint, G. R. W. 1983 Leaf damage in tropical rain forest canopies. In *Tropical rain forest: ecology and management* (ed. S. L. Sutton, T. C. Whitmore & A. C. Chadwick), pp. 229–239. Oxford: Blackwell Scientific Publications.

P. D. MOORE (*Division of Biosphere Sciences, King's College London, U.K.*). Dr Collinson has emphasized the importance of mammals in determining the course of fruit evolution in the angiosperms. Is it not possible that bird vectors may have played just as important a part in the evolution of dispersal mechanisms and hence in fruit structure?

M. E. COLLINSON. It is indeed possible that bird vectors played an important role in the evolution of fruit dispersal. Unfortunately, birds generally have a poorer fossil record than mammals and, except at certain sites, articulated skeletons are very rare. Nevertheless, there is direct evidence of frugivory in birds as early as the Middle Eocene from gut contents at Messel in Germany (see, for example, Schaal & Ziegler (1988), figures 82 and 217). This is clearly an area ripe for future study.

J. OLLERTON (*Oxford Polytechnic, U.K.*). Apologies for being an ignorant plant ecologist, but what exactly does Dr Collinson mean by the term 'scansorial'?

R. McN. ALEXANDER (*Department of Pure and Applied Biology, University of Leeds, U.K.*). I have an uncomfortable feeling that if I were given a selection of mammalian postcranial skeletons I might have difficulty in judging which belonged to small ground-living mammals, which to scansorial mammals and which to arboreal ones. Has Dr Collinson clear criteria for such distinctions? A multivariate analysis of skeletal form might make it easier to formulate reliable criteria.

M. E. COLLINSON. The criteria for recognizing tree-dwelling mammals relate to the necessity to rotate the feet to appropriate positions on the uneven substrate and to grip surfaces that depart significantly from the horizontal. Degree of rotation of the hind foot for instance can be judged from the distribution of articular facets on the astragalus and calcaneum (ankle joint). Gripping broad tree surfaces (e.g. trunks) is effected by strong laterally compressed claws (covering similarly shaped terminal phalanges) and describes the scansorial habit. Gripping narrow branches describes the arboreal habit and is effected by foot or tail prehensility, which can be assessed from the distribution of articular facets on the foot bones or from length of the tail plus degree of development of transverse processes and haemal arches on the tail vertebrae. A 'small' ground (semiterrestrial) mammal is one which lacks both these specializations and those typifying the 'large' ground (fully terrestrial) mammal, such as hooves, reduction of digits, or wrist and ankle joints that restrict lateral movement. It is thus the most difficult to recognize, but is a widespread locomotor mode today. We agree that a more rigorous classification of morphotypes relating to these locomotor modes is desirable.

How important is biotic pollination and dispersal to the success of the angiosperms?

JEREMY J. MIDGLEY[1] AND WILLIAM J. BOND[2]

[1] *CSIR – Division of Forest Science and Technology, Saasveld F.R.C., Private Bag x6515, George, 6530, South Africa*
[2] *Botany Department, University of Cape Town, Private Bag Rondebosch, 7700, South Africa*

SUMMARY

The rise to dominance of the angiosperms has frequently been hypothesized to be due to reproductive innovations, especially those involving coevolution with biotic gene dispersers. Wind pollination is considered inefficient in comparison with insect pollination. The scope for speciation is considered to be greater when biotic gene dispersers are involved. The biogeographical restriction of extant conifers to stressful environments is also considered to be due to the inefficiency of wind pollination in species-rich environments.

We consider both fossil and contemporary evidence and conclude that biotic gene dispersal has not been as important as have innovations affecting growth rates in explaining the rise of angiosperms. We conclude that differences in growth rates, especially in the regeneration phase, can explain the rise of the angiosperms and the extant biogeography of gymnosperms.

1. INTRODUCTION

From the time they first appeared there was no stopping the angiosperms. Starting at the beginning of the Cretaceous at low palaeo-latitudes or at the mid-Cretaceous at high palaeo-latitudes there was a steady exponential increase in the number of angiosperm species at any particular location (Crane & Lidgard 1989). Today the angiosperms completely dominate the vegetation of low latitudes. However, gymnosperms still survive in high latitudes, high altitudes and nutrient-poor soils. What accounts for the surge of angiosperms in the past and their distribution in the present?

Competitive advantage has been invoked to explain the replacement sequence in the fossil record of all plant groups up to but excluding the angiosperms (Knoll 1984, 1986). Chaloner & Sheerin (1981) suggest that a change from an open 'r-selected' environment to a closed 'K-selected' forest might have led to reproductive changes from the homosporous to heterosporous condition because large propagules would be favoured in these more competitive, deeply shaded environments. Thus Knoll (1984) invoked improved architecture and conductance for why the rhyniophytes were replaced by the trimerophytes in the Devonian and similarly for why they in turn were replaced by progymnosperms and pteridophytes. Knoll (1984) argued that because of the life style of plants they, unlike animals, would not be prone to mass extinctions. Their seeds and buds would allow them to survive a short-term catastrophe like a bolide impaction. He argued that other factors such as progressive climatic change and vegetative competition determined replacement sequences in the plant fossil record. How-

ever, Knoll (1984) and others (Regal 1977; Raven 1977) reasoned that for the angiosperms, superior reproductive innovations explained their rise to dominance. The same reasoning has been invoked to explain the contemporary distribution of conifers (Regal 1977).

The supposedly superior reproductive innovations of angiosperms can be put into three classes. First those that favour increased levels of gene dispersal per unit effort. Faithful pollinators may have moved pollen further as well as facilitating a decrease in pollen production and thus of the costs of reproduction. Similarly vertebrate dispersers may have resulted in longer distance seed dispersal and to safer sites. Second are those attributes that facilitate increased levels of mate choice. Various incompatibility systems and gametophytic competition within the style allowed more careful screening of partners. Third are those attributes that favour increased rates of speciation or decreased rates of extinction. Coevolution with pollinators or dispersers, pollinator-mediated isolating mechanisms and dispersal away from pathogens and predators on or near parents may have lead to increased speciation rates or decreased extinction rates.

These classes may or may not be mutually exclusive. For example coevolution between plants and animals could increase levels both of gene flow and speciation. Alternately increased levels of gene flow could reduce the chances of isolation and thus decrease speciation rates. Paradoxically both better (Regal 1977; Raven 1977) and poorer gene flow (Niklas *et al.* 1983; Doyle & Donoghue 1986) of the angiosperms has been invoked to explain their success. There is very little information as to how greater levels of heterozygosity (a by-product of efficient gene dispersal), mate choice or speciation could lead to evolutionary success. For example there is no strong relation between heterozygosity and tree performance (Charlesworth & Charlesworth 1987) or geographic range (Ledig 1986).

Here we critically review arguments invoking the importance of coevolution with animal pollinators and dispersers in angiosperm success. We concur with Stebbins (1981) that traits associated with rapid growth and reproduction were far more important in angiosperm success and their contemporary distribution.

There are many measures of evolutionary success including diversity of a clade, longevity of a clade, its geographic spread and ecological abundance (Stebbins 1981; Doyle & Donoghue 1986; Wilson 1987; Bond 1989). There need be no necessary association between these different criteria. Speciose taxa may contribute a small percentage of the biomass of an area (e.g. Orchidaceae; *Ficus*) whereas single taxa may be ecologically important over large areas (e.g. *Pinus sylvestris*, *P. resinosa*; in Africa, *Colophospermum mopane*). Until recently, available reviews of fossil data provided numbers of species and little about abundance (Niklas *et al.* 1983), although this can be a misleading criterion of success. For example, there are now more species of fern than ever before although they are in a relative state of decline. We assess the importance of traits both for diversity and ecological dominance and their possible interdependence.

2. THE IMPORTANCE OF BIOTIC GENE DISPERSAL

(a) *Pollination*

Regal (1977, 1982), Raven (1977), Crepet (1983, 1984) and others have argued that angiosperm success can be attributed to biotic pollination. Wind pollination, characteristic of most gymnosperms, is traditionally considered to be an inefficient wasteful process relying on the excessive production of pollen (Whitehead 1983). Pollen flow is considered to be especially limited in a closed canopy environment. Therefore, anemophilic tree species should be restricted to species-poor environments where pollen movement between conspecifics would not be intercepted by other species. This argument is invoked to explain the almost total absence of anemophiles (angiosperm or gymnosperm) from diverse lowland tropical forests and their restriction to stressed (= species-poor) environments (Regal 1977). Regal (1977) suggested that the introduction of angiosperm canopies into essentially wind-pollinated Mesozoic communities diluted the efficiency of wind pollination and began the downward spiral of non-angiosperms.

(b) *Dispersal*

Regal (1977) argued that when bird and mammal seed vectors evolved they could carry large numbers of seeds into new environments and that these sites were also beyond the distribution of specialized seed predators. This, combined with insect pollination, allowed plants to produce out-crossed offspring in populations of widely dispersed individuals: precisely the conditions where wind pollination would be unreliable. Regal (1977) considered that many of the families of important insect pollinators long predated the origins of angiosperms whereas the timing of the origins of avian dispersers was more closely synchronized. However, fleshy 'fruits' are not particular to angiosperms and are common in many gymnosperms. Thus he reasoned that it was the combined effect of both types of biotic gene dispersal that was crucial in explaining the success of angiosperms.

(c) *Criticisms*

Midgley & Bond (1989, 1991) and Bond (1989) have questioned the biotic gene-dispersal hypothesis on many grounds. Pollination arguments can be challenged on the basis of changing understanding of wind pollination, genetic evidence and the timing of appearance of biotic partners in the fossil record.

Recent studies challenge the generalization that wind pollination is inefficient. In the Knysna Forest, Midgley (1989) found mature seeds on female individuals of the extant conifer *Podocarpus falcatus* that were 70 m away from the nearest reproductive male. This indicates that pollen can move considerable distance in closed canopy forests. Niklas (1985) has indicated that wind pollination is not as random a process as once conceived. Direct compilations of limited data on neighbourhood area suggest that wind-pollinated species have the largest gene flow (Crawford 1984). Ellstrand *et al.* (1990) reviewed the available data for genetic structure of gymnosperms and concluded that they generally have high diversity but low spatial differentiation. On the other hand the cycad they investigated had high spatial variation and low genetic diversity. New studies of gene movement show that the commonly held paradigm that gene flow is very limited (Ehrlich & Raven 1969) is not strictly accurate for either wind- or insect-pollinated plants (Devlin & Ellstrand 1990; Ellstrand & Marshall 1985) and the whole field needs review. This has important implications for those hypotheses for angiosperm success which assumed that biotic gene dispersal increased neighbourhood size over that of wind-pollinated and dispersed taxa. However at this stage there is no evidence that non-angiosperms have low genetic diversity or that they are characterized by low gene flow.

Contemporary biogeographic evidence for the restriction of gymnosperms to species-poor forests is weak. Some conifers are rare in forests (*Torreya*, *Podocarpus falcatus*) and southern hemisphere forests with podocarps can be as rich in species as forests without gymnosperms (Midgley & Bond 1989). Also it is hard to imagine how shrub-like early angiosperms (Crane 1987) diluted pollen movement of canopy gymnosperms in Cretaceous forests.

There is increasing evidence for pre-angiosperm insect pollination which also weakens the reproductive coevolution argument. Biotic gene dispersal is not an angiosperm apomorphy. Angiosperms are part of an anthophyte clade (Doyle & Donoghue 1986; Crane 1985) with flower-like structures some of which show direct evidence of pollination by insects (for example, the Bennettitales) long before the appearance of angiosperms (Crepet *et al.*, this symposium). Field studies indicate that contemporary cycads are also insect pollinated, and animal dispersed (Norstog *et al.* 1986). The fossil record of some of these insects dates back to the Jurassic (J. Donaldson, personal communication, 1991). Thus the evolution of insect-pollinated plants may have long preceded the appearance of angiosperms.

Crepet (1984) has argued that constancy of insect pollinators was a key to angiosperm success. The most constant pollinators are among the Apoidea, and Lepidoptera but these only evolved in the late Cretaceous and early Tertiary (Crepet 1984). Crepet explained the discrepancy by pointing to the supposed slow initial angiosperm diversification. However, recent compilations of palynological data suggests that the origins of angiosperms and their radiation were exponential (Crane & Lidgard 1989) so that insect radiation may have followed rather than preceded initial angiosperm radiation. We believe this casts serious doubt on any need to invoke the concomitant evolution of pollinators or dispersers for the success of angiosperms.

Coevolution arguments find no benefits in wind pollination. Yet wind pollination evolved very early in angiosperm history (Humphries 1988) and still predominates in large, successful families including graminoid monocots (Gramineae, Cyperaceae, Juncaceae, Restionaceae) and woody dicots (Fagaceae). It has evolved repeatedly from insect-pollinated ancestors at the level of families, genera (*Leucadendron*, *Acer*) and even species (*Plantago*, *Salix*). Anemophily is a neglected field relative to biotic pollination (most citations are pre-1960). There are few explanations (besides scarcity of insects, for example, Crepet (1981)) offered for its repeated evolution from insect-pollinated ancestors. The bottom line is that we have no answer for questions like 'what is the difference for the plant whether it is wind, bird or insect pollinated?' What is worse is that there is no real paradigm to take on such important evolutionary questions.

Vertebrate dispersal of fruits probably also had little to do with angiosperm success (Tiffney 1984, 1986; Herrera 1989). Vertebrate-dispersed fruits occur in many gymnosperms (*Taxus*, *Torreya*, *Podocarpus*, *Pinus pinea*, etc.), including extinct Jurassic forms and the insect-pollinated cycads (Tiffney 1986; Herrera 1989). Early angiosperms were abiotically dispersed and fleshy fruits (and appropriate frugivorous birds) only became common in the Tertiary, long after the initial angiosperm radiation (Tiffney 1984, 1986). In summary, the combination of long-distance pollination and long-distance biotic dispersal proposed by Regal (1977) did not exist during the Cretaceous radiation of the angiosperms.

3. THE IMPORTANCE OF MATE CHOICE

This hypothesis argues first that reproductive innovations of angiosperms result in rapid selection and, second, implies that this generates diversity ('plas-ticity') that accounts for angiosperm success (Mulcahy 1979). Mulcahy (1979) contended that pollen in gymnosperms arrives singly with the first arrival having a 'head start'. In insect-pollinated angiosperms, in contrast, pollen would arrive *en masse* and simultaneously. Pollen competition on the stigma and style might then result in more vigorous, competitively superior offspring (Mulcahy 1979). Animal pollination, in combination with the stigma and style arising from carpel closure, should thus speed up natural selection giving 'the angiosperms an evolutionary plasticity that fuelled their dramatic rise to dominance in the world flora' (Mulcahy 1979).

There are very few direct studies of microgametophytes per ovule to test Mulcahy's arguments. From the limited data that exists, ovules of wind-pollinated plants actually have more microgametophytes from which to choose (table 1). In *Ginkgo*, several male gametophytes may occur in the pollen chamber, each producing two spermatozoids, resulting in competition for fertilization at the two archegonia (Favre-Duchartre 1958). Indeed *Ginkgo*'s flagellated sperm has been interpreted as a device to win the competitive race across the pollination chamber (Haig & Westoby 1989). There are also few studies on the simultaneity of pollen deposition in wind-pollinated plants. The long delay between pollination and fertilization in wind-pollinated gymnosperms and some angiosperms (e.g. *Quercus*) has been interpreted as a mechanism for sampling more pollen for gametophytic selection (Willson & Burley 1983). If wind-pollinated plants use long pre-fertilization times to sample pollen mates, then pollination mode may influence rates of reproduction more than the diversity of pollen parents. Clearly more work is needed on wind-pollinated plants to test Mulcahy's assumptions of small microgametophyte or ovule ratios and 'first come first served'.

Mulcahy's argument ignores the significance of multiple paternity as a mechanism for generating genetic diversity. Many zoophilous taxa disperse pollen in polyads as opposed to the monads of anemophiles. Because of this and the higher probability of pollen from different donors arriving at wind-pollinated flowers, Kress (1981) argued that the potential for multiple paternity is actually highest in wind-pollinated taxa. Thus Kress (1981) argued that the low numbers of ovules per ovary which characterize wind-pollinated taxa may be the product of potentially higher levels of kin selection.

It has long been recognized that angiosperms have a high potential for discarding unwanted genotypes without proportionate loss of reproductive capacity (Buchholz 1922). However the relative importance of pre- versus post-zygotic selection in either controlling mate choice or costs of reproduction is still poorly understood. While pre-zygotic choice through pollen competition may be important, the evidence is still controversial (Snow & Mazer 1988; Charlesworth 1988; Stephenson *et al.* 1988). Mate choice does exist in gymnosperms (e.g. polyembryony, delayed fertilization (Wilson and Burley 1983); selective ovule abortion (Haig & Westoby 1989) but pre-zygotic mechanisms

Table 1. *The size of microgametophyte populations per ovule and pollination mode*

(Data from Snow (1986), *Epilobium*; Levin (1990), *Phlox* and *Polemonium*; M. Honig (personal communication, 1990), *Staberoha* (Cape fynbos Restionaceae).)

species	pollination	microgametophytes per ovule
Epilobium canum	insect	2
Phlox drummondii	insect	4.7
Polemonium viscosum	insect	3
Staberoha stokoei	wind	5.2

are poorly developed. Gymnosperms may be selective, but their selection may cost them dearly in terms of reproductive rates (see, for example, Stebbins (1974); Bond (1989)).

In summary, critical evidence is lacking for the importance of biotic pollination in Mulcahy's mate choice hypothesis for the generation of angiosperm diversity. However, other structural features (the style, tectate pollen, reduced embryo sac, double fertilization, etc.) certainly do influence the capacity for mate choice without risking reproductive capacity. It is still not at all clear, though, whether the undoubted plasticity of angiosperm form springs from mate choice (Mulcahy 1979) or physiological attributes regulating organ growth such as rates of solute supply, the fine network of leaf veins, novel intercalary meristematic tissue (Halle *et al.* 1978; Bond 1989) or some combination. Plants have a remarkable capacity for macromutational change (Gottlieb 1984). The implication is that growth form diversity may depend more on regulation and flexibility of growth processes than soft selection acting on multiple alleles.

4. THE IMPORTANCE OF SPECIATION RATES

Could biotic gene dispersal increase speciation rates and would speciation rates (there being no doubt about the superior speciation rates of angiosperms) alone explain the rise to dominance of the angiosperms? Crepet (1984) and Crepet & Friis (1987) linked angiosperm diversification with the appearance of constant pollinators (Apoidea and Lepidoptera) in the late Cretaceous and early Tertiary. He argued that Coleoptera and Diptera, which had been present throughout the Mesozoic, 'do not have the ethologies associated with heightened promotion of speciation events' (Crepet 1984). However Coleoptera are the major pollinators of the highly speciose Asteraceae. In the rich Cape flora of South Africa, there is no strong evidence that speciation depends on the pollinator type and even less to suggest that speciation rates are a product of coevolution with constant pollinators (Midgley & Bond 1991). Contrary to Crepet's generalization, important pollinators of the Cape flora are Coleoptera, Diptera and birds (Johnson 1991).

Many wind- and insect-pollinated angiosperm lineages are almost equally speciose (Midgley & Bond 1991). It would be instructive to compare geographic differences in the scale of speciation between wind-pollinated and zoophilous taxa. A preliminary assessment of the Cape flora indicates little difference. Thus centres of endemism of important wind-pollinated (Restionaceae) and (mainly) animal-pollinated taxa (Proteaceae, Ericaceae, Rutaceae, Fabaceae) coincide (Oliver *et al.* 1983). Using the data in Oliver *et al.* (1983) species of Restionaceae occupy a mean of 4.5 quarter degree grid squares. Mean numbers of quarter degree squares occupied by species of the Proteaceae, Ericaceae, Penaeaceae and Bruniaceae (all zoophilous taxa) are all more than for the Restionaceae, and the Diosmeae (Rutaceae) are less. In other words the distribution patterns of zoophilous taxa does not indicate limited distributions. Finally, despite there being very few species with fleshy fruits, the Cape flora is exceptionally species rich (Bond & Goldblatt 1984). Our impression is that pollination syndromes (and fleshy fruits, for example, see Herrera (1989)) are not correlated with speciation rates or the sizes of distributions.

Coevolution thus appears less important than a series of morphological traits which promote speciation in angiosperms whether they are animal or wind pollinated or dispersed. Closure of the carpel, development of a stigma, style, and tectate pollen all lead to development of sporophytic and gametophytic reproductive barriers (Burger 1981; Zavada & Taylor 1986). Tectate pollen is absent in gymnosperms with the exception of an extinct group of conifers, the Cheirolepidiaceae (Alvin 1982; Watson 1988). Cheirolepids included extremely diverse growth forms ranging from stem succulents to cypress-like trees (Watson 1988). As in the angiosperms, this diversity has been attributed to the potential benefits of enhanced mate choice (Alvin 1982). It remains to be seen whether diversification of these conifers was also associated with biotic, rather than wind, pollination (Watson 1988).

5. ALTERNATIVES: A COMPETITION HYPOTHESIS

We believe that the importance of reproductive superiority has been over-emphasized whereas rapid growth rates and their attendant ecological effects, based on vegetative and reproductive structures (Bond 1989), have been neglected.

The capacity for rapid vegetative growth and fast reproduction allows angiosperms to colonize disturbed areas and rapidly occupy space: Grime's (1979) ruderal and competitive strategies respectively. In productive environments, angiosperms can outcompete gymnosperm seedlings, outgrow gymnosperm canopies in forests, and reproduce earlier than gymnosperms to survive disturbance such as fire (Bond 1989).

Angiosperms have been able to diversify into a wide variety of growth forms (annuals, broad leaves), architecture (*sensu* Halle *et al.* 1978) and regeneration strategies (rhizomes) which gymnosperms have not. This plasticity can be attributed to vegetative traits such as the more efficient transport system (see, for example, Halle *et al.* (1978); Bond (1989)) and a different meristematic organization (Klekowski *et al.* 1985). Many conifers lack regular axillary meristems (Burrows 1987), which may severely constrain flexibility of tree architecture.

The problem of the peculiar biogeography of the gymnosperms and their persistence over large areas is not addressed by either the mate choice or speciation hypotheses. However the biogeography of non-angiosperms is interpretable in terms of their retreat into environments where fast-growing angiosperms are restrained (Bond 1989). The absence of conifers in tropical lowland forests has been explained in terms of the limitations of wind pollination in species-rich communities (Regal 1977). It seems to us an equally plausible solution is that conifers cannot successfully compete in these productive environments. Rundel *et al.* (1991) have used stable isotopes to compare water-use efficiency between angiosperms and gymnosperms. Semi-arid gymnosperms have water-use efficiencies comparable to woody angiosperms. In this environment the low evaporative flux of gymnosperms is not a handicap. However tropical conifers depart markedly from angiosperms with lower water-use efficiency, consistent with lower maximum conductance values and reduced carbon uptake. Rundel *et al.* (1991) conclude that their results are consistent with the growth-rates hypothesis (based on low conductive efficiency) and help explain the scarcity of conifers in the tropics.

Wolfe (1990) showed strong correlation of leaf physiognomy and climate in most extant vegetation types. Tropical angiosperm forests have large leaves. Arborescent conifers may be excluded from these environments simply because they do not have large leaves.

Rapid reproductive rates in angiosperms has been attributed to the reduced embryo, double fertilization, acceleration of pollen tube growth by the style and the lower cost of insect pollination (Stebbins 1976, 1981). There still are no clear data on the costs of reproduction for different pollination syndromes, i.e. whether the insect pollination of the angiosperms represented a saving. Anemophiles produce vast quantities of pollen and this has traditionally been taken to indicate its inefficiency and high cost. An alternate explanation for this increase in pollen production is that it is due to intramale competition (Lloyd 1984; Bond & Midgley 1988), rather than evidence of inefficiency. Zoophilous taxa gain no advantage in producing more than a certain amount of pollen because their vectors get saturated. This trade-off does not apply to wind-pollinated taxa. As far as the costs are concerned, the pollen of anemophiles is energetically cheaper to produce than that of zoophilous species (Petanidou & Vokou 1990). Further work is needed to clear up the overall costs of reproduction of wind versus insect pollination.

(a) *Implications of the growth-rates hypothesis*

We believe the growth-rates hypothesis can be expanded to include other groups besides the conifers.

Ferns for instance should be found in dark nutrient-poor and cold environments or in growth forms that angiosperms cannot exploit. Thus in productive environments we hypothesize that angiosperm epiphytes (e.g. Brommeliaceae, Orchidaceae) exclude pteridophytic epiphytes. We expect parallels in distribution patterns between ferns and conifers despite their different 'pollination systems'. Cycad biogeography remains unsynthesized. We suggest competition in regeneration niche, especially from grasses, as crucial. We expect them to be vulnerable to frequent disturbance (e.g. fire) and to be more common in stressful (micro)sites (e.g. nutrient-poor, shallow soils and shady). We predict that non-angiosperms (e.g. Bennetitaleans) persisted longer on nutrient-poor (arenaceous) than on nutrient-rich (argillaceous) substrates.

The growth-rates hypothesis is consistent with the fossil record with its steady-state take over, rather than a bumpy pattern correlated with the evolution of other animal lineages. According to the growth-rates hypothesis, the take-over by angiosperms is expected to have taken place first in productive environments (low palaeolatitudes). The rise of the angiosperms may have been facilitated by other factors such as advanced pollinators and vertebrate dispersers but was not dependent on them.

Finally, growing information on other physical and biological attributes of Cretaceous environments are consistent with the growth-rates hypothesis but are not addressed by the coevolution hypotheses. High oxygen levels during the Cretaceous may have led to more frequent fires (Robinson 1989) and changes in browsing levels of the dinosaur fauna (Bakker 1978), both favoured rapid growth and short generation time. Abiotic changes would also have promoted the necessary fast growth. Cretaceous climates were warm and latitudinal gradients were very weak, especially near the poles (Spicer & Chapman 1990). High oxygen levels would have produced a more oxidative soil environment rather than a reducing podzolic soil environment produced by cool wet gymnosperm dominated flora (Robinson 1989). The break-up of Gondwanaland in an erosional rather than depositional cycle would have favoured angiosperms by providing nutrient-rich substrates.

6. CONCLUSIONS

We have taken a particularly critical stance against arguments involving the importance of biotic gene dispersal. We do so to emphasize the weak evidence for several widely accepted hypotheses and the need to explain the many exceptions of diverse and successful taxa without animal pollination or dispersal.

What are the problems with the fast growth-rate hypothesis? First, although it can explain the great diversity of growth forms among the angiosperms, it has nothing to say on speciation within a growth form. Secondly, although the early success of the ephedroids (see Crane & Lidgard 1989) is compatible with the growth-rates hypothesis it does not explain their subsequent failure to match the angiosperms. Gnetalean vascular morphology may not produce the transport efficiency of modern angiosperms but we lack data. Finally the impact of environmental changes, especially if they are abrupt, needs further attention.

W.J.B. acknowledges financial support from the University of Cape Town and the Foundation for Research and Development.

REFERENCES

Alvin, K. L. 1982 Cheirolepidiacae: biology, structure and paleoecology. *Rev. Palaeobot. Palynol.* **37**, 71–98.

Bakker, R. T. 1978 Dinosaur feeding behaviour and the origin of flowering plants. *Nature, Lond.* **274**, 661–663.

Bond, P. & Goldblatt, P. 1984 Plants of the Cape flora. *J. S. Afr. Bot.* (Suppl.) **13**, 1–455.

Bond, W. J. & Midgley, J. J. 1988 Allometry and sexual differences in leaf size. *Am. Nat.* **131**, 901–910.

Bond, W. J. 1989 The tortoise and the hare: ecology and angiosperm dominance and gymnosperm persistence. *Biol. J. Linn. Soc.* **36**, 227–249.

Bucholz, J. T. 1922 Volumetric studies of seeds, endosperms and embryos in *Pinus ponderosa* during embryonic differentiation. *Bot. Gaz.* **108**, 232–244.

Burrows, G. E. 1987 Leaf axil anatomy in the Araucariaceae. *Aust. J. Bot.* **35**, 631–640.

Chaloner, W. G. & Sheerin, A. 1981 The evolution of reproductive strategies in early land plants. In *Evolution today* (ed. G. G. E. Scudder & J. L. Reveal), pp. 93–100. Proc. 2nd International Congress of Systematic and Evolutionary Biology.

Charlesworth, D. 1988 Evidence for pollen competition in plants and its relationship to progeny fitness: a comment. *Am. Nat.* **132**, 298–302.

Charlesworth, D. & Charlesworth, B. 1987 Inbreeding depression and its evolutionary consequences. *A. Rev. Ecol. Syst.* **18**, 237–268.

Crane, P. R. 1985 Phytogenetic analysis of seed plants and the origin of the angiosperms. *Ann. Missouri Bot. Gard.* **72**, 716–793.

Crane, P. R. 1987 Vegetational consequences of the angiosperm diversification. In *The origins of angiosperms and their biological consequences*, (ed. E. M. Friis, W. G. Chaloner & P. R. Crane) pp. 107–145. Cambridge University Press.

Crane, P. R. & Lidgard, S. 1989 Angiosperm diversification and paleolatitudinal Gradients in Cretaceous Floristic Diversity. *Science, Wash.* **246**, 675–678.

Crawford, T. J. 1984 What is a population? In *Evolutionary ecology* (ed. B. Shorrocks), pp. 135–173. Oxford: Blackwell Scientific Publications.

Crepet, W. L. 1981 The status of certain families of the Amentiferae during Middle Eocene and some hypotheses regarding the evolution of wind pollination in dicotyledonous angiosperms. In *Palaeobotony, palaeoecology and evolution* (ed. K. J. Niklas), pp. 103–128.

Crepet, W. L. 1983 The role of insect pollination in the evolution of the angiosperms. In *Pollination biology* (ed. L. Real), pp. 29–50. Orlando, Florida: Academic Press.

Crepet, W. L. 1984 Advanced (constant) insect pollination mechanisms: pattern of evolution and implications vis-a-vis angiosperm diversity. *Ann. Missouri Bot. Gard.* **71**, 607–630.

Crepet, W. L. & Friis, E. M. 1987 The evolution of insect pollination in angiosperms. In The origins of angiosperms and their biological consequences (ed. E. M. Friis, W. G. Chaloner & P. R. Crane), pp. 181–199. Cambridge University Press.

Devlin, B. & Ellstrand, N. C. 1990 The Development and application of a refined method for estimating gene flow from angiosperm paternity analysis. *Evolution* **44** (2), 248–259.

Doyle, J. A. & Donoghue, M. J. 1986 Seed plant phylogeny and the origin of angiosperms: an experimental cladistic approach. *Bot. Rev.* **52**, 321–402.

Ehrlich, P. R. & Raven, P. H. 1969 Differentiation of populations. *Science, Wash.* **165**, 1228–1232.

Ellstrand, N. C. & Marshall, D. L. 1985 Interpopulation gene flow by pollen in wild radish *Raphanus sativus. Am. Nat.* **126**, 606–616.

Ellstrand, N. C., Ornduff, R. & Clegg, J. M. 1990 Genetic structure of the Australian Cycad, *Macrozamia communis* (Zamiaceae). *Am. J. Bot.* **77** (5), 677–681.

Favre-Duchartre, M. 1958 *Ginkgo*, an oviparous plant. *Phytomorphology* **8**, 377–390.

Gottlieb, L. D. 1984 Genetics and morphological evolution in plants. *Am. Nat.* **123**, 681–709.

Grime, J. P. 1979 *Plant strategies and vegetation processes.* Chichester: John Wiley and Sons.

Haig, D. & Westoby, M. 1989 Selective forces in the emergence of the seed habit. *Biol. J. Linn. Soc.* **38**, 215–238.

Halle, F., Oldeman, R. A. & Tomlinson, P. 1978 *Tropical trees and forests: an architectural analysis.* New York: Springer-Verlag.

Herrera, C. M. 1989 Seed dispersal by animals: a role in angiosperm diversification? *Am. Nat.* **133**, 309–322.

Humphries, C. 1988 Patterns before process. *Nature, Lond.* **333**, 300–301.

Johnson, S. 1991 Plant-animal relationships in the Cape fynbos. In *Cape fynbos* (ed. R. M. Cowling). Oxford University Press. (In the press.)

Klekowski, E. J., Kazarinova-Fukshansky, N. & Mohr, H. 1985 Apical meristems and mutation: stratified meristems and angiosperm evolution. *Am. J. Bot.* **72**, 1788–1800.

Knoll, A. H. 1984 Patterns of extinction in fossil record of vascular plants. In *Extinctions* (ed. M. Nitecki), pp. 21–67. University of Chicago Press.

Knoll, A. H. 1986 Patterns of change in plant communities through geological time. In *Community ecology* (ed. J. Diamond & T. Case), pp. 126–141. New York: Harper & Row.

Kress, W. J. 1981 Sibling interactions and the evolution of pollen unit, ovule number and pollen vector in angiosperms. *Syst. Bot.* **6**, 101–112.

Ledig, F. T. 1986 Heterozygosity, heterosis and fitness in outbreeding plants. In *Conservation biology* (ed. M. Soule), pp. 77–104. Massachusetts: Sinauer.

Levin, D. A. 1990 Sizes of natural microgametophyte populations in pistils of *Phlox drummondi. Am. J. Bot.* **77**, 356–363.

Lloyd, D. G. 1984 Gender allocations in outcrossing plants. In *Perspectives on plant population ecology* (ed. R. Dirzo & J. Sarukhan), pp. 277–300. Massachusetts: Sinauer.

Midgley, J. J. & Bond, W. J. 1989 Evidence from southern African Coniferales for the historical decline of the gymnosperms. *S. Afr. J. Sci.* **85**, 81–84.

Midgley, J. J. & Bond, W. J. 1991 Ecological aspects of the rise of the angiosperms: a challenge to the reproductive superiority hypothesis. *Biol. J. Linn. Soc.* (In the press.)

Midgley, J. J. 1989 Pollen dispersal distances for a conifer canopy species in the Knysna Forest. *S. Afr. J. Bot.* **55** (6), 662–663.

Mulcahy, D. H. 1979 The rise of the angiosperms: a genecological factor. *Science, Wash.* **206**, 20–3.

Norstog, K. J., Stevenson, D. W. & Niklas, K. J. 1986 The

role of beetles in the pollination of *Zamia furfuracea* L. fil. (Zamiaceae). *Biotropica* **18**, 300–306.

Niklas, K. J. 1985 The aerodynamics of wind pollination. *Bot. Rev.* **51** (3), 328–386.

Niklas, K. J., Tiffney, B. H. & Knoll, A. H. 1983 Patterns in vascular plant diversification. *Nature, Lond.* **303**, 614–616.

Oliver, E. G. H., Linder, H. P. & Rourke, J. P. 1983 Geographical distribution of present-day Cape taxa and their phytogeographical significance. *Bothalia* **14**, 427–440.

Petanidou, T. & Volkou, D. 1990 Pollination and pollen energetics in Mediterranean ecosystems. *Am. J. Bot.* **77** (8), 986–992.

Raven, P. H. 1977 A suggestion concerning the Cretaceous rise to dominance of the angiosperms. *Evolution* **31**, 451–452.

Regal, P. J. 1977 Ecology and evolution of flowering plant dominance. *Science, Wash.* **196**, 622–629.

Regal, P. J. 1982 Pollination by wind and animals: ecology of geographic patterns. *A. Rev. Ecol. Syst.* **13**, 497–524.

Robinson, J. M. 1989 Phanerozoic O_2 variation, fire, and terrestrial ecology. *Palaeogeogr. Palaeoclimatol. Palaeoecol.* (Global and Planetary Change Section) **75**, 223–240.

Rundel, P. W., Goldstein, G. H., Orosco, A. & Kang, S. J. Foliar carbon isotope ratios in conifers: evidence for constraints of water use efficiency on ecological success. (In preparation.)

Snow, A. 1986 Pollination dynamics in *Epilobium canum* (Onagraceae): consequences for gametophytic selection. *Am. J. Bot.* **73**, 139–151.

Snow, A. & Mazer, S. J. 1988 Gametophytic selection in *Raphanus raphanistrum*: a test for heritable variation in pollen competitive ability. *Evolution* **42**, 1065–1075.

Spicer, R. A. & Chapman, J. L. 1990 Climate change and the evolution of high-latitude terrestrial vegetation and floras. *Trends Ecol. Evol.* **5**, 279–284.

Stebbins, G. L. 1974 *Flowering plants: evolution above the species level*. Cambridge, Belknap: Harvard University Press.

Stebbins, G. L. 1981 Why are there so many species of flowering plants? *Bioscience* **31**, 573–577.

Stephenson, A. G., Winsor, J. A., Schlichting, C. D. & Davis, L. E. 1988 Pollen competition, nonrandom fertilization and progeny fitness: a reply to Charlesworth. *Am. Nat.* **132**, 303–308.

Tiffney, B. H. 1984 Seed size, dispersal syndromes and the rise of angiosperms: evidence and hypothesis. *Ann. Missouri Bot. Gard.* **71**, 551–576.

Tiffney, B. H. 1986 Evolution of seed dispersal syndromes according to the fossil record. In *Seed dispersal* (ed. D. R. Murray), pp. 273–305. Sydney: Academic Press.

Watson, J. 1988 The Cheirolepidiaceae. In *Origin and evolution of gymnosperms* (ed. C. Beck), pp. 382–447. New York: Columbia University Press.

Whitehead, D. R. 1983 Wind pollination; some ecological and evolutionary perspectives. In *Pollination biology* (ed. L. Real), pp. 97–108. Academic Press.

Wilson, E. O. 1987 Causes of ecological success: the case of the ants. *J. Anim. Ecol.* **56**, 1–9.

Wilson, M. F. & Burley, N. 1983 *Mate choice in plants*. Princeton University Press.

Wolfe, J. A. 1990 Palaeobotanical evidence for a marked temperature increase following the Cretaceous/Tertiary boundary. *Nature, Lond.* **343**, 153–156.

Zavada, M. S. & Taylor, T. N. 1986 The role of self-incompatibility and sexual selection in the gymnosperm-angiosperm: a hypothesis. *Am. Nat.* **128**, 538–550.

Discussion

E. A. Jarzembowski (*Booth Museum of Natural History, Brighton, U.K.*). In the search for animal vectors in biotic pollination, it is tempting to generalize about insects, although they greatly outnumber vertebrates and angiosperms with diversities in major orders such as Diptera, Coleoptera and Hymenoptera exceeding 10^5 species. It is certainly true that these orders had appeared before the Cretaceous radiation of the angiosperms, but closer examination of the fossil data indicates a greater turnover of families at the Lower–Upper Cretaceous boundary than at the Cretaceous–Tertiary boundary. The radiation of angiosperms seem to have been more disruptive than the K/T catastrophe although the same data shows an increase in overall insect diversity during the Cretaceous implying insect–plant interaction.

Reference

Jarzembowski, E. A. 1989 Cretaceous insect extinction. *Mesozoic Res.* **2** (1), 25–28.

J. J. Midgley. The problem is whether the radiation of associated animals (mainly insects and birds), especially pollinators and dispersers but also herbivores, preceded or followed that of angiosperms. We have argued that plant–animal interactions were less important than plant–plant interactions. We expect the radiation of the angiosperms to have preceded that of associated animal groups.

P. J. Grubb (*Botany School, University of Cambridge, U.K.*). Gymnosperms do not dominate equally effectively all kinds of vegetation in which plants grow slowly. They commonly dominate in cold areas, on soils poor in nutrients, and in areas as dry as those with mediterranean climates but they are generally absent from semi-deserts, with the exception of *Welwitschia* and some *Ephedra* species. Most gymnosperms depend on maintaining several cohorts of long-lived leaves on the adult, which is not a viable option for shrubs that need to survive very long and severe droughts (often shedding most or all of their leaves) and then grow quickly when rain comes. *Welwitschia* and *Ephedra* effectively parallel succulents.

J. J. Midgley. We agree with Dr Grubb's comments. In general gymnosperms appear to be relatively ineffective, especially in the regeneration niche, in any ecosystem where speed (rapid growth) rather than endurance is crucial. Also, present gymnosperms are morphologically fairly uniform. We need further information as to why so few seem to have converged on the succulence niche.

Abiotic pollination: an evolutionary escape for animal-pollinated angiosperms

PAUL ALAN COX

Department of Botany & Range Science, Brigham Young University, Provo, Utah 84602, U.S.A.

SUMMARY

Early botanists considered abiotic pollination to be primitive in angiosperms. But we now deduce from studies of palaeoecology and of extant 'primitive' angiosperms that animal pollination was concomitant with the rise of the angiosperms. Recent studies of wind and water pollination in angiosperms also show these systems to be highly sophisticated. If entomophily contributed to the rise of the early angiosperms, why should many of their descendants have later evolved abiotic pollination systems?

Although entomophily was initially advantageous to the early angiosperms, abiotic pollination systems may be superior in areas of low species diversity, newly colonized habitats, and places with extremely short growing seasons or other adverse climatic conditions. Abiotically pollinated plants are not constrained by the range of animal pollinators, and as a result are spectacularly successful in long-distance dispersal. Abiotic pollination also offers an escape from deleterious sexual selection and from dependency on pollinators that are climatically limited in their distribution in space or time and vulnerable to extinction. Because evolution of abiotic pollination frequently leads to dicliny or dichogamy, it is largely irreversible. This evolutionary irreversibility coupled with lowered rates of extinction and speciation give wind- or water-pollinated taxa unique phylogenetic profiles.

> As a large quantity of pollen is wasted by anemophilous plants, it is surprising that so many vigorous species of this kind abounding with individuals should still exist in any part of the world; for if they had been rendered entomophilous, their pollen would have been transported by the aid of the senses and appetites of insects with incomparably greater safety than by the wind...It seems at first sight a still more surprising fact that plants, after having been once rendered entomophilous, should ever have again become anemophilous.
>
> (Darwin 1876, p. 407)

1. INTRODUCTION: IS ENTOMOPHILY A PRIMITIVE CHARACTER IN ANGIOSPERMS?

In *The effects of cross and self fertilization in the vegetable kingdom*, Charles Darwin struggled with the implausibility of entomophilous plants ever undergoing retrograde evolution to a more primitive condition: anemophily. Yet even though Darwin knew of a few probable cases of reversion such as *Poterium sanguisorba* (Rosaceae), in his time this mystery was not regarded as particularly pressing as only a few species were believed to have so reverted.

This lack of interest in the evolution of abiotic pollination systems in angiosperms in the late 19th century is hardly surprising: at the time, abiotic pollination was unquestionably regarded as primitive. In the evolutionary scheme proposed by Axell (1869) wind-pollinated species were considered to be the most imperfect and primitive phanerogams because of their immense waste of pollen. Delpino (1868, 1869) added an even lower rung to the evolutionary ladder: hydrophily, a somewhat surprising denigration given the previous biological and theological significance

accorded to water pollination (Darwin 1790–1791; Paley 1802; Cox 1988a).

This belief in the inherent primitiveness of abiotic pollination was promulgated by Müller (1873) in his influential *Die Befruchtung der Blumen durch Insekten*:

> Finally from the wind-pollinated angiosperms, entomophilous flowers arose, as insects came first accidentally and afterwards regularly to seek their food on flowers, and as natural selection fostered and perfected every change which favored insect visits and thereby aided cross-fertilization...The transition from wind pollination to insect pollination and the first traces of adaptation to insects, could only be due to the influence of quite short-lipped insects with feebly developed color sense.
>
> (Müller 1881, pp. 593–594)

That these hierarchical schemes were regarded as orthodox was underscored by Darwin's laudatory preface to the English version of Müller (Müller 1883). Müller's ranking of pollination types was later adopted by Verhoeff (1894) and Knuth (1898).

Today, this mystery is far greater than in Darwin's time, for recent research has shown that the early angiosperms were not wind pollinated, as Darwin and his contemporaries assumed, but instead insect pollinated. This contemporary view comes not only from pollination biology (Meeuse 1961; 1984; Proctor & Yeo 1973; Fægri & Van der Pijl 1979; Thien *et al.* 1985) but from paleobotany and systematics (Doyle 1978; Crane *et al.* 1986; Crepet & Friis 1987; Doyle & Donoghue 1987).

Key to this understanding has been evidence that the first angiosperms and their sister groups were probably entomophilous. Friis & Crepet (1987) argue that the hermaphroditic reproductive structures of the Bennettitalean genera *Williamsoniella* and *Wielandiella* were functionally analogous to a magnolialean flower, with beetles and possibly dipterans as the putative pollinators. Pollen evidence, however, is more equivocal (Retallack & Dilcher 1981).

Friis & Crepet (1987) propose *Illicium* (Illiaceae–Magnoliidae) pollination by flies and beetles (Thien *et al.* 1983) as a modern analogue of early angiosperm pollination. Other possible modern analogues include *Zygogynum* (Winteraceae), *Drimys* (Winteraceae) and *Degeneria* (Degeneriaceae), all of which are pollinated by flies and beetles (Thien 1980; Thien *et al.* 1985; Miller 1989).

Although reconstruction of ancient pollination ecologies from fossils and modern analogues is admittedly uncertain, there now appears to be little argument concerning the primitiveness of entomophily in angiosperms, although some of the early angiosperms may have utilized both insects and the wind as pollen vectors (Dilcher 1979; Retallack & Dilcher 1981). This conclusion, if true, dramatically increases the difficulty of Darwin's mystery: the evolution of abiotic pollination must now be explained in not only a few specialized species, but rather in all anemophilous and hydrophilous angiosperms.

2. POSSIBLE TRANSITIONS BETWEEN ANEMOPHILY AND ENTOMOPHILY

What are some of the possible reasons that anemophily might evolve in an entomophilous species? This question can be divided into two separate issues: (i) the possible evolutionary pathways by which abiotic pollination could be derived from entomophily, and (ii) the evolutionary forces that could favour such a transition.

Obvious candidates for such a change are species that use a combination of biotic and abiotic pollen vectors. In these species strict entomophily and anemophily appear to be but two endpoints on a spectrum, with some species using both vectors sequentially whereas other species use them simultaneously. For example, the flowers of *Bartsia* and *Lathraea* (Scrophulariaceae) and many Ericaceae, such as *Calluna vulgaris* and *Erica carnea*, when first opened, have their pollen dispersed only by insects. Later, though, when the filaments elongate and the anthers are exerted beyond the corolla mouth, the pollen is carried by the wind to other flowers (von Marilaun 1895). A similar mechanism exists in *Cyclamen* with the

oily pollenkitt which serves to adhere the pollen to insects reportedly vanishing as the flowers progress in age (Hilldebrand 1897).

Plants with simultaneous entomophily and anemophily are also known. Many species of the otherwise anemophilous genus *Luzula* (Juncaceae) such as *L. lutea*, *L. nivea* and *L. lactea* produce brilliantly coloured flowers with brightly coloured pseudo-nectaries; in these species pollen is dispersed by both insects and the wind (Buchenau 1871; Knuth 1898). Insects also visit otherwise anemophilous taxa such as *Plantago lanceolata* and *P. media* (Plantaginaceae), species of the genus *Tilia* (Tiliaceae) and a nectarless species of *Helianthemum* (Cistaceae) (Clifford 1962; Fægri & van der Pijl 1973; Proctor & Yeo 1973; Primack 1978).

Tilia is worthy of further note as Pohl (1937 *b*) found that *Tilia* has a pollen:ovule ratio well within the range of entomophilous species. Yet the intensity of *Tilia* pollen rain throughout Britain is very high, with up to 83.6 grains cm^{-3} $year^{-1}$ in Cambridge (Hyde 1950). The British pollen rain is also rich in other predominantly entomophilous species such as *Salix* and *Sambucus*.

In some genera there is variation among the species in the relative allocation to anemophily or entomophily. For example *Acer* (Aceraceae) species differ in pollenkitt, and have corresponding differences in pollination syndrome (Hesse 1979). *Acer negundo* (Aceraceae) is strictly anemophilous, *A. campestre*, *A. opalus*, and *A. pseudoplatanus* use a combination of pollination vectors, whereas *A. platanoides* is entomophilous. Similar variation occurs in *Thalictrum* (Ranunculaceae). Most species are visited by insects, but *T. polygonum* and *T. flavum* use both insects and the wind as pollen vectors, whereas *T. dioicum* and *T. minor* are strictly anemophilous, bearing traces of their entomophilous ancestry (Müller 1883; Proctor & Yeo 1973; Melampy & Hayworth 1980).

It appears, therefore, that several entomophilous species have either some component of anemophilous pollination or are potentially anemophilous, even if their wind-dispersed pollen is not efficacious in fertilization. Inefficacy of wind-borne pollen probably indicates a female floral or stigmatic morphology that is aerodynamically inappropriate for pollen reception. The evolution of anemophily in such taxa would therefore require development of appropriate aerodynamic surfaces for pollen capture.

Are there cases where the converse is true, i.e. where the aerodynamic surfaces for pollen capture are present, but where the adaptations necessary for wind-borne pollen have not yet evolved? One such example occurs in the Pandanaceae, which has tristichous phyllotaxy as a synapomorphy uniting the family. In *Pandanus* this phyllotactic pattern functions to position the floral bracts beneath the pistillate inflorescence in such a way as to impart the aerodynamic characteristics necessary for pollen capture by the stigmatic surfaces (Cox 1985, 1990).

However, similar flow patterns are also produced by the tristichous bracts of pistillate inflorescences of the vertebrate-pollinated species *Freycinetia reineckei* (Pandanaceae). But in the case of *Freycinetia*, tristichous phyllotaxis appears as a preadaption to anemophily because its pollen is covered with a sticky, lipid-rich pollenkitt, which effectively prevents wind dispersal. Thus the tristichous arrangement of floral bracts in *Freycinetia* could prove functional for wind pollination only with a loss of pollenkitt and some changes, such as lengthening of the internode, in the staminate inflorescence. None of these changes would be dependent on major alteration in structural gene sequences, but would require only subtle difference in the timing of regulatory gene activity. A process similar to this may have allowed wind pollination in *Pandanus* to evolve from vertebrate-pollinated ancestors (Cox 1990).

Similar evolutionary lability between anemophily and entomophily as a result of developmental timing is apparent in the mangrove family Rhizophoraceae. *Ceriops decandra*, which is insect pollinated, and *Rhizophora mangle*, which is wind pollinated but has relic nectaries, have strikingly similar floral ontogenies (Juncosa & Tomlinson 1987). Judging from the Pandanaceae and the Rhizophoraceae, the evolutionary gap between anemophily and entomophily, measured in the amount of necessary structural genetic change, may be small indeed. This realization leads to a new question: are all plant species equally likely to evolve abiotic pollination systems, even if these systems have selective advantage?

It seems unlikely that a species with a sympetalous corolla and anthers basifixed to the corolla throat could easily evolve anemophily because of inherent difficulty in pollen dispersal. But a species with exerted stamens, and short or no corolla would be a more likely candidate. Fægri & van der Pijl (1973) suggest that just such a reduction of tepals in brush blossoms can facilitate anemophily and offer a putative transition series in *Thalictrum*. In hydrophilous species, floral appendages can play an important role unrelated to their ancestral function of pollinator attraction. In both *Vallisneria* and *Enhalus* (Hydrocharitaceae) the staminate flowers are carried on the water surface by the reflexed petals, and fall into the depressions created by the long, floating petals of the pistillate flowers (Cox 1988a). The staminate flowers of the related seagrass genus *Halophila* still bear long petals underwater, although these do not appear to be functional in pollination.

3. IS ENTOMOPHILY ALWAYS BENEFICIAL?

The idea that abiotic pollination systems are inherently primitive is no longer tenable: recent research has highlighted the sophisticated mechanisms used for pollen dispersal and capture in anemophilous and hydrophilous angiosperms (Whitehead 1983; Niklas & Buchmann 1985; Cox 1988a). For example Typhaceae, which used to be considered the most primitive monocotyledonous family (Engler & Gilg 1924), has a wind-pollination system that is remarkably sophisticated and successful. Wind-dispersed tetrads of *Typha latifolia* pollen frequently grow pollen tubes from the stigma of one flower to the stigma of an adjacent flower (Nicholls & Cook 1986). Resultant seed is set

high; although in the field only one third of the stigmas are pollinated, remarkably over two thirds of the uniovulate flowers set seed (Krattinger 1978). Similarly, the seagrass genera *Amphibolis* and *Thalassodendron*, once believed to be the most primitive seagrasses (Hartog 1970), have recently been found to have intricate pollination mechanism involving tidally synchronized release of submarine male flowers, rapid expulsion of noodle-like pollen on the water surface, and formation of floating search vehicles of high efficiency (Cox & Knox 1989; Cox 1991).

At the community level, abiotic pollination systems do not appear to be as inefficient or wasteful as might have been previously assumed. In this symposium, Midgley & Bond report that wind-pollinated taxa are as speciose as animal-pollinated taxa in the Cape flora, and do not appear to be at a reproductive disadvantage. The number of cases in which anemophilous species clearly have entomophilous ancestry also argues against a global disadvantage for wind-pollinated species. Under what conditions, then, might anemophily be advantageous?

To answer, it is necessary to consider the conditions under which insect pollination is disadvantageous. This may seem an odd question, as it has been argued by many authors (and summarized in this symposium by Crepet *et al.*) that insects and insect pollination have played a pivotal role in the evolution of angiosperms, with many of the reproductive structures of angiosperms evolving in response to insect pollination. Yet perhaps evolutionary importance is not always synonymous with evolutionary beneficence.

Is it possible, for example, that tightly coevolved plant–pollinator relations such as those involving figs and fig wasps, or yuccas and yucca moths, may not represent an evolutionary achievement, but rather an evolutionary dead-end?

4. POSSIBLE MECHANISMS DRIVING THE EVOLUTION OF ABIOTIC POLLINATION SYSTEMS FROM ENTOMOPHILOUS SYSTEMS

Under what conditions could anemophily possibly evolve from entomophily? In an ecological sense, if we consider a plant such as a *Tilia* or *Plantago* species which uses both biotic and abiotic vectors for pollination, we might ask under what conditions would it be advantageous for individuals to decrease the proportion of pollen dispersed by animal vectors, and to increase the proportion of pollen dispersed by abiotic vectors. In an evolutionary sense we can examine the conditions that might cause anemophily to become an evolutionary stable strategy (ESS) such that a wind-pollinated population would be invulnerable to invasion by an insect pollinated mutant.

The following are examples of circumstances that could favour abiotic pollination as an ESS.

(a) *Range expansion or colonization ability*

Raven (1977) has argued that entomophily allowed the angiosperms to persist as widely dispersed populations, particularly in tropical ecosystems where accurate pollination transfer between widely spaced individuals is important. However this advantage is likely to hold only at intermediate scales of distance, and rapidly vanishes in the case of long-distance dispersal, where the migration ability of an entomophilous plant is ultimately constrained by the range of its pollinators.

Even on a local scale, entomophilous plants that colonize microsites unattractive to pollinators may have reduced success (Watson 1969; Handel 1983). But clearer distinctions emerge on large geographical scales, because differences in long-distance dispersal ability and colonization appear to be closely tied to modes of pollination. *Pandanus* (Pandanaceae), with its anemophilous pollination system, can colonize any appropriate island regardless of the island's faunal composition. This anemophily, coupled with facultative apomixis, has given it a range throughout the Pacific and Indian oceans that is much larger than the related vertebrate pollinated genus *Freycinetia*, and vastly larger than the related entomophilous genus *Sararanga* (Cox 1985, 1990).

The facilitation of long-distance dispersal by abiotic pollination has been very important in seagrasses. Even though only 12 genera exist, there are few coastlines or islands with appropriate substrates that have not been successfully colonized by these hydrophilous monocotyledons. Of the temperate genera, *Zostera* (Zosteraceae) has a strikingly large distribution, being found along the coasts of southeastern Africa, Europe, Australia, New Zealand, eastern Asia, and both coasts of North America. A tropical genus, *Halophila*, is found throughout the Indian Ocean, South Pacific, Caribbean, and west tropical Atlantic (den Hartog 1970). Although the seagrasses are capable of clonal growth, recent research has highlighted their high levels of sexual reproduction and resultant genetic diversity (Cox 1988a; Cox *et al.* 1991). Dependency on specific animal-pollination vectors would have greatly limited the ability of both Pandanaceae and the seagrasses to colonize new areas, particularly oceanic islands.

(b) *Pollinator extinction*

The spectre of extinction frequently haunts plant species dependent on highly specific pollinators: if the pollinator goes extinct, their own survival is threatened. However, confirmed reports of plant extinctions linked to pollinator losses are few, because the pollination biologies of rare species are little studied and the extinction processes are likely rapid. In Hawaii, the extinction of indigenous avian pollinators imperiled *Freycinetia arborea* until a new pollinator was inadvertently introduced (Cox 1983). In the western United States the overwhelming majority of rare or endangered plants are entomophilous while very few are anemophilous (Harper 1979); Clearly anemophilous taxa are rarely forced to face the loss of their pollen vector. However, although I suggest that extinction rates should be lower in anemophilous taxa, Eriksson & Bremer (1991) find that the R value (speciation rate

minus the extinction rate) is significantly higher for animal-pollinated than for abiotically pollinated species, a result similar to that of Niklas *et al.* (1985). Although this may constitute a contradiction to my hypothesis, these higher R values may merely reflect the higher speciation rates of entomophilous taxa.

(c) *Deleterious sexual selection*

Under sexual selection, small increases in male reproductive effort can result in disproportionately large increases in reproductive success (Willson 1979), far in excess of what could be obtained through investment in abiotic pollination (figure 1). However, sexual selection can rapidly drive phenotypic evolution to the point where even large investments in male reproductive functions result in only marginal increments of reproductive success.

Consider, for example, pollination in the Madagascar orchid *Angraecum*. In 1862 Darwin predicted that *A. sesquipedale*, which has a 22 cm long spur concealing nectar, would be found to be pollinated by a giant hawkmoth with a 22 cm long proboscis. Just such a hawkmoth, *Xanthopan morgani praedicta*, was discovered over 40 years later (Rothschild & Jordan 1903). The existence of these long corolla spurs (up to 30 cm in *A. sesquipedale* and *A. sororium*) and interminable moth proboscii, may in fact, represent a case of coevolution out of control. Nilsson (1985, 1987) suggests a process of remorseless selection for increasing tongue length in the moths, because even slightly shorter-proboscis moths are unable to reach the nectar in the longest spurs. But the flowers also have remorseless selection for increasing spur length, as flowers with slightly shorter spurs are unable adequately to position their pollinaria on the moths with the longest proboscis. As result, each round of selection requires the orchid to pay a higher price for pollination.

Sexual selection can therefore alter relative advantages of abiotic and biotic pollination by changing the expected return on male reproductive effort (figure 1). As sexual selection advances, abiotic pollination provides a more favourable return on male reproductive investment, although in many cases (such as *Angraecum*) switching to abiotic pollination may be morphologically impossible.

(d) *Low community diversity*

In Utah and Idaho, dominant, overstory plant species are typically anemophilous, whereas species with low frequency and dominance values are usually entomophilous (Ostler & Harper 1978). This prevalence of anemophilous species in temperate regions, and their relative absence from tropical regions, has long been noted (Whitehead 1983). Clearly the structural and microclimatic features of moist tropical forests may impede wind pollination, whereas those of the wind-swept arid regions may favour it. But I suggest that the tremendous differences in species diversity between the tropics and temperate areas may also play an important role.

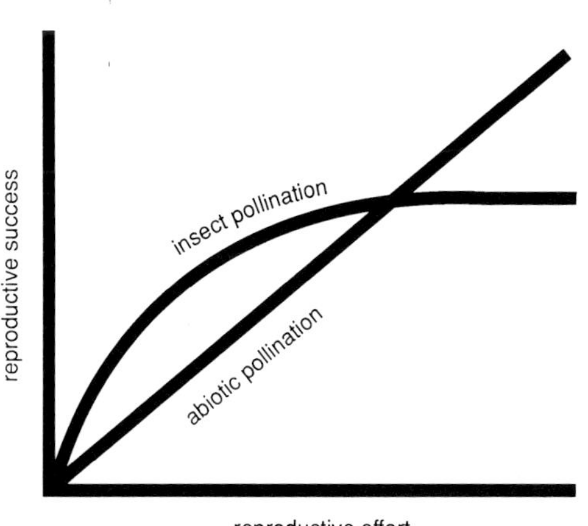

Figure 1. Effects of sexual selection.

Even though anemophilous species can aerodynamically partition the wind, selectively capturing their own grains based on size and velocity (Buchman *et al.* 1989), there might be a limit to the number of species that can effectively divide the wind between themselves. Furthermore, as species diversity increases, population densities must of necessity decrease, also lowering the proportion of self-pollen landing on stigmas. But although some theoretical upper limit on the number of wind-pollinated species in a community may exist, there probably is a lower limit to the number of entomophilous species that can occupy a plant community. For unless an entomophilous species has an extremely smeared phenology, it is in a sense dependent on other flowering species that maintain its pollinators when it itself is not in flower. Thus in communities of extremely low species diversity abiotic pollination systems may be favoured.

(e) *Outbreeding in founder populations*

Even though most wind-borne pollen travels short distances, a small amount is dispersed further away (Whitehead 1983). Even long-distance pollination is possible, with significant quantities of *Nothofagus* (Fagaceae) pollen travelling over 5000 km from South America to Tristan da Cuhna in the middle of the Atlantic (Proctor & Yeo 1973). Although the extremely short viability of hydrophilous pollen, measured in hours, makes fertilization at long distances unlikely, the much longer viability of anemophilous pollen makes long-distance pollination, however rare, a distinct possibility. Such long-distance pollination may be important in maintaining a modicum of seed set in small isolated populations of self-incompatible, dichogamous, or diclinous populations at the extension of the species range, and may also play an important genetic role (see §4*h*).

(f) *Phenological and developmental flexibility*

Another reason that anemophilous taxa are successful in temperate areas may be that they are not phenologically constrained in the same manner that entomophilous taxa are. Selection drives flowering

times of entomophilous taxa to be closely tied to periods of maximal pollinator activity, in many cases effectively excluding early season and late-season flowering. Highly synchronous phenologies may also be more difficult to achieve in entomophilous taxa either because of the need to avoid pollinator satiation or because of the necessity to maintain trap-lining pollinators. Hence in entomophilous taxa there may be potential conflicts between flowering times optimized for pollination success and flowering times optimized for fruit development.

Lack of dependency on biotic vectors may, conversely, allow anemophilous taxa to have longer developmental periods for fruits, to have highly synchronous flowering bouts when favoured by climatic conditions, or even to have a highly smeared flowering phenology if bet hedging is in order. The ability precisely to calibrate flowering with fruiting times may give anemophilous taxa a tremendous advantage in areas with short growing seasons or adverse climates.

(g) *Phenological synchrony with abiotic phenomena*

As mentioned above, abiotic pollination systems might allow for better synchrony with abiotic phenomena. This capability may be of importance in fine-tuning flowering times for solar-linked phenomena, such as the change of seasons in temperate and arctic climates. But even in these areas there is a rough correlation between onset of pollinator activity and change of the solar calendar. No such correlation exists between pollinator activity and lunar-linked phenomena such as tides, and the ability to synchronize flowering with tides is crucial in the seagrasses. For many seagrasses, surface pollination is possible only during low spring tides and flowering times are tightly synchronized with these periods (Cox 1988*a*, 1991; Cox & Knox 1989). Even submarine-pollinated plants show phenologies synchronized with the tides (Cox & Tomlinson 1988; Cox *et al.* 1990).

(h) *Changed population structure*

Abiotically pollinated plants are more likely to be able to persist as pioneering populations in new habitats because they are not constrained by pollinator ranges. They can also set seed and maintain limited genetic contact with parent populations as a result of long-distance pollination. However, except for freshwater hydrophilous plants growing in ponds, small pioneering populations of entomophilous plants are more likely to be reproductively isolated, leading to higher speciation rates. The ability of anemophilous species to undergo dramatic range expansions, persisting as groups of geographically dispersed subpopulations that still maintain limited genetic contact with another, dramatically reduces the likelihood of extinction due to local environmental perturbations. This maintains a broad ecological amplitude for anemophilous species, allowing results of successful genetic experiments to be rapidly communicated throughout the range of the species.

(i) *Phenotypic plasticity*

The increased ecological amplitude of anemophilous plants allows them to have greater colonization ability, particularly in early successional stages. Thus rather than being micro-site specialists, anemophilous species can utilize more generalized, *r*-type strategies. This inherent phenotypic plasticity allows anemophilous species to adapt more rapidly to changing abiotic environments. The indiscriminate nature of the wind also may give anemophilous species more opportunities for interspecific and even intergeneric hybridization. Anemophily may give better chances for autopolyploid experiments to succeed, particularly in the colonization of newly opened habitats such as in the genus *Atriplex* (Chenopodiaceae), where autopolyploids invaded new habitats opened with the rapid Pleistocene recession of Lake Bonneville (Stutz & Sanderson 1979).

5. ANEMOPHILY IN ANGIOSPERMS: A CASE OF IRREVERSIBLE EVOLUTION?

Crepet (1981) has hypothesized that wind pollination in angiosperms may have evolved in seasonally dry tropical environments before the mid-cretaceous before the evolution of the bees (Willemstein 1987). Competition for pollinators in beetle-pollinated taxa favoured the evolution of wind pollination, with these anemophilous tropical taxa later invading more temperate areas. But if this is the case, why does abiotic pollination appear to be a one-way evolutionary street? Why have not anemophilous taxa reverted to entomophily with changing conditions?

Much of the reason, I suspect, has to do with the strong need for anemophilous plants to separate male and female sexual functions in either time or space. Entomophilous hermaphrodites use the same floral displays, odours, and pollinator rewards to disperse as well as to attract pollen. But in wind- or water-pollinated taxa, the mechanisms for pollen dispersal tend to interfere with the mechanisms for pollen capture, and hence dimorphic flowers tend to predominate. For example, 75% of seagrass genera are dioecious, and this separation of the sexes is likely maintained because of potential mechanical interference between the sexes (Cox 1988*a*, *b*). The frequency of dicliny is also high in wind-pollinated taxa. Even those abiotically pollinated species with monomorphic flowers tend to be strongly dichogamous, effectively separating male and female functions in time.

This strong selection for physical separation of male and female reproductive functions in abiotically pollinated taxa may lead to what Bull & Charnov (1985) term 'irreversible evolution'. Their concept is similar to the evolutionary 'blind alleys' and 'traps' articulated by Simpson (1953). In short, although anemophily can often rapidly evolve in an entomophilous taxon, the evolution of entomophily in an anemophilous taxon is very rare, particularly in dioecious taxa.

For entomophily to be an ESS the population must be vulnerable to invasion by an entomophilous mutant. Let us consider the case of a dimorphic anemophilous population invaded by an entomophilous mutant; i.e. a mutant plant that suddenly starts attracting insects through the production of nectar, showy floral displays, or other contrivances. If the mutant is male, it must sire more ovules than anemophilous males for the invasion to succeed. However, because the other females in the population lack pollinator attractants, insect-dispersed pollen would be at a disadvantage compared to wind-dispersed pollen, as female plants already have the aerodynamic mechanisms necessary for pollen capture. On the other hand, if the mutant is female and effective at attracting insects, the insect visitors would not be likely to bring pollen from the unattractive, nectarless males unless they were pollen robbers. In either case it is difficult to imagine what would attract the insects to the nectarless flowers of the other sex, and hence pollen flux would probably be far lower for the entomophilous males or females. However, for the invasion to succeed, male progeny of the mutant must successfully disperse more pollen than anemophilous males, and female progeny must set more seed than anemophilous females, all in the absence of nectar rewards. Hence the interference between male and female functions in anemophilous taxa ultimately leads to dicliny or dichogamy, making the evolution of anemophily largely irreversible.

6. CONCLUSIONS

Such cases, however, are intelligible, as almost all plants require to be occasionally intercrossed; and if any entomophilous species ceased to be visited by insects, it would probably perish unless it were rendered anemophilous…The same result would follow to a certainty, if winged insects ceased to exist in any district, or became very rare.   (Darwin 1876, pp. 407–408)

In dealing with the few known cases of evolutionary 'reversion' to anemophily Darwin suggested one possible factor favouring anemophily: the vulnerability of entomophilous species to pollinator extinction. Darwin's list has here been modestly expanded, but these few emendations scarcely seem adequate to deal with the current problem. Moreover, much like Gödel's theorem, the technique of analysis used in this essay, relying heavily on ecological correlation, points to its own inadequacy. If speciation and extinction rates are themselves skewed by abiotic pollination, then species correlations are themselves invalid (Cox 1990). What is needed is painstaking phylogenetic analysis to determine the precise clades in which abiotic pollination evolved. Although a few attempts have been made along these lines (Cox 1990), a broader phylogenetic approach is needed to reveal the underlying patterns.

REFERENCES

Axell, S. 1869 *Om anordningarna för fanergogama växternas befuktning*. Stockholm.

Buchenau, F. 1871 Beobachtungen über die Bestäubung von *Juncus bufonius*, L. *Bot. Zeit.* **19**, 845–852.

Buchmann, S. L., O'Rourke M. K. & Niklas, K. J. 1989 Aerodynamics of *Ephedra trifurca* III. Selective pollen capture by pollination droplets. *Bot. Gaz.* **150**, 122–131.

Bull, J. J. & Charnov, E. L. 1985 On irreversible evolution. *Evolution* **39**, 1149–1155.

Clifford, H. T. 1962 Insect pollinators of *Plantago lanceolata* L. *Nature, Lond.* **193**, 196.

Cox, P. A. 1983 Extinction of the Hawaiian avifauna resulted in a change of pollinators for the ieie, *Freycinetia arborea. Oikos* **41**, 195–199.

Cox, P. A. 1985 Islands and dioecism: insights from the reproductive ecology of *Pandanus tectorius* in Polynesia. In *Studies on Plant Demography: a Festschrift for John L. Harper* (ed. J. White), pp. 359–372. London: Academic Press.

Cox, P. A. 1988*a* Hydrophilous pollination. *A. Rev. Ecol. Syst.* **19**, 261–280.

Cox, P. A. 1988*b* Monomorphic and dimorphic sexual strategies: a modular approach. In *Plant reproductive ecology: strategies and patterns* (ed. J. Lovett Doust & L. Lovett Doust), pp. 80–97. Oxford University Press.

Cox, P. A. 1990 Pollination and the evolution of breeding systems in Pandanaceae. *Ann. Missouri Bot. Gard.* **77**, 816–840.

Cox, P. A. 1991 Hydrophilous pollination of a dioecious seagrass, *Thalassodendron ciliatum* (Cymodoceaceae) in Kenya. *Biotropica* **23**, 159–165.

Cox, P. A., Elmqvist, T. & Tomlinson, P. B. 1990 Submarine pollination and reproductive morphology in *Syringodium filiforme* (Cymodoceaceae). *Biotropica* **22**, 259–265.

Cox, P. A. & Knox, R. B. 1989 Two-dimensional pollination in hydrophilous plants: convergent evolution in the genera *Halodule* (Cymodoceaceae), *Halophila* (Hydrocharitaceae), *Ruppia* (Ruppiaceae), and *Lepilaena* (Zannichelliaceae). *Am. J. Bot.* **76**, 164–175.

Cox, P. A., Laushman, R. & M. Rucklehaus. 1991 Surface and Submarine pollination in the seagrass *Zostera marina* L. *Bot. J. Linn. Soc.* (In the press.).

Cox, P. A. & Tomlinson, P. B. 1988 The pollination ecology of a Caribbean seagrass, *Thalassia testudinum* (Hydrocharitaceae). *Am. J. Bot.* **75**, 958–965.

Crepet, W. L. 1981 The status of certain families of the Amentiferae during the middle Eocene and some hypotheses regarding the evolution of wind pollination in dicotyledonous angiosperms. In *Paleobotany, paleoecology, and evolution*, vol. 2 (ed. K. J. Niklas), pp. 27–77. New York: Praeger.

Crane, P. R., Friis, E. M. & Pedersen, K. R. 1986 Lower Cretaceaous angiosperm flowers: fossil evidence on early radiation of dicotyledons. *Science, Wash.* **232**, 852–854.

Crepet, W. L. & Friis, E. M. 1987 The evolution of insect pollination in angiosperms. In *The origins of angiosperms and their biological consequences* (ed. E. M. Friis, W. G. Chaloner & P. R. Crane), pp. 181–201. Cambridge University Press.

Darwin, C. 1876 *The effects of cross and self fertilization in the vegetable kingdom.* London: John Murray.

Darwin, E. 1790–1791 *The botanic garden, a poem. In two parts. Part II. The loves of the plants.* London: J. Johnson.

Delpino, F. 1868–1869 Ulteriori osservazioni e considerazioni sulla dicogamia nel regno vegetale. *Atti Soc. Ital. Sci. Nat., Milano* **11**, 265–332; **12**, 21–141, 179–233.

Dilcher, D. L. 1979 Early angiosperm reproduction: an introductory report. *Rev. Palaeobot. Palynol.* **27**, 291–328.

Doyle, J. A. 1978 Origin of angiosperms. *A. Rev. Ecol. Syst.* **9**, 365–392.

Doyle, J. A. & Donoghue, M. J. 1987 The origins of angiosperms: a cladistic approach. In The origins of angiosperms and their biological consequences (ed. E. M. Friis, W. G. Chaloner & P. R. Crane), pp. 17–49. Cambridge University Press.

Engler, A. & Gilg, E. 1924 *Syllabus der Pflanzenfamilien.* Berlin: Verlag von Grebuder Borntreager.

Fægri, K. & van der Pijl, L. 1979 *The principles of pollination ecology*, 3rd revised edn. Oxford: Pergamon Press.

Handel, S. N. 1983 Pollination ecology, plant population structure, and gene flow. In *Pollination biology* (ed. L. Real), pp. 163–211. Orlando, Florida: Academic Press.

Harper, K. T. 1979 Some reproductive and life history characteristics of rare plants and implications of management. *Great Basin Natur. Mem.* **3**, 129–137.

Hartog, C. den. 1970 *The sea-grasses of the world. (Verh. K. ned. Akad. Wet.* **59**). Amsterdam: North Holland.

Hildebrand, F. 1897 Über die Bestäubung bei den Cyclamen-Arten. *Ber. dt. bot. Ges.* **15**, 292–298.

Hesse, M. 1979 Ultrastructure and distribution of pollenkitt in the insect and wind pollinated genus *Acer* (Aceraceae). *Pl. Syst. Evol.* **131**, 277–290.

Hyde, H. A. 1950 Studies in atmospheric pollen. IV. Pollen deposition in Great Britain. *New Phytol.* **49**, 393–420.

Juncosa, A. M. & Tomlinson, P. B. 1987 Floral development in mangrove Rhizophoraceae. *Am. J. Bot* **74**, 1263–1279.

Kerner von Marilaun, A. 1895 *The natural history of plants, vol II* (transl. F. W. Oliver). London: Gersham.

Knuth, P. 1898 *Handbuch der Blütenbiologie.* Band I–III. Leipzig: Verlag von Wilhelm Engelmann.

Krattinger, K. 1978 Biosystematische Untersuchungen innerhalb der Gattung *Typha* L. Dissertation, University of Zürich.

Meeuse, B. J. D. 1961 *The story of pollination.* New York: Ronald Press.

Meeuse, B. J. D. & Morris, S. 1984 *The sex life of flowers.* New York: Facts on File.

Melampy, M. N. & Hayworth, A. M. 1980 Seed production and pollen vectors in several nectarless plants. *Evolution* **34**, 1144–1154.

Miller, J. M. 1989 The archaic flowering plant family Degeneriaceae its bearing on an old enigma. *Natn. Geogr. Res.* **5**, 218–231.

Müller, H. 1873 *Die Befruchtung der blumen durch Insekten.* Leipzig.

Müller, H. 1883 *The fertilization of flowers* (transl. D. W. Thompson). London: MacMillan.

Nicholls, M. S. & Cook, C. D. K. 1986 The function of pollen tetrads in *Typha* (Typhaceae). *Veröff. geobot. Inst., Zürich* **87**, 112–119.

Niklas, K. J. & Buchmann, S. L. 1985 Aerodynamics of wind pollination in *Simmondsia chinensis. Am. J. Bot.* **72**, 530–539.

Nilsson, L. A. 1985 Monophily and pollination mechanisms in *Angraecum arachnites* Schltr. (Orchidaceae) in a guild of long-tongued hawk-moths (Sphingidae) in Madagascar. *Biol. J. Linn. Soc.* **26**, 1–19.

Nilsson, L. A. 1987 On the evolution of flower deepness by hawkmoths. XIV *Int. Bot. Cong.* (Berl.). Progr. 4–20, p. 193.

Ostler, W. K. & Harper, K. T. 1978 Floral ecology in relation to plant species diversity in the Wasatch mountains of Utah and Idaho. *Ecology* **59**, 848–861.

Paley, W. 1802 *Natural theology, or evidences of the existence and attributes of the deity collected from the appearances of nature.* London: R. Faulder.

Primack, R. B. 1978 Evolutionary aspects of wind pollination in the genus Plantago (Plantaginaceae). *New Phytol.* **81**, 449–458.

Proctor, M. & Yeo, P. 1973 *The pollination of flowers*. London: Collins.

Raven, P. H. 1977 A suggestion concerning the rise to dominance of the angiosperms. *Evolution* **31**, 451–452.

Retallack, G. & Dilcher, D. L. 1981 A coastal hypothesis for the dispersal and rise to dominance of flowering plants. In *Paleobotany, paleoecology, and evolution*, vol. 2 (ed. K. J. Niklas), pp. 27–77. New York: Praeger Publishers.

Rothschild, L. W. & Jordan, K. 1903 A revision of the lepidopterous family Sphingidae. *Novit. zool.* **9**, 1–972.

Simpson, G. G. 1953 *The major features of evolution*. New York: Columbia University Press.

Stutz, H. C. & Sanderson, S. C. 1979 The role of polyploidy in the evolution of *Artriplex canescens*. In *Arid land plant resources* (ed. J. R. Goodin & D. K. Northington), unpaginated. Lubbock, Texas: Texas Tech University.

Taylor, T. N. 1981 Pollen and pollen organ evolution in early seed plants. In *Paleobotany, Paleoecology, and Evolution*, vol. 2 (ed. K. J. Niklas), pp. 1–25. New York: Praeger Publishers.

Thien, L. B. 1985 Patterns of pollination in the primitive angiosperms. *Biotropica* **12**, 1–13.

Thien, L. B., Bernhardt, P., Gibbs, G. W., Pellmyr, O., Bergsrtom, G., Groyh, I. & McPherson, G. 1985 The pollination of *Zygogynum* (Winteraceae) by a moth, *Sabatinca* (Micropterigidae): an ancient association? *Science*, Wash. **227**, 540–543.

Thien, L. B., White, D. A. & Yatsu, L. Y. 1983 The reproductive biology of a relict – *Illicium floridanum* Eliis. *Am. J. Bot.* **70**, 719–727.

Verhoeff, C. 1894 Blumen und Insekten der Insel Norderney. *Nova Acta Leopoldina* **61**, 45–216.

Watson, P. S. 1969 Evolution in closely adjacent plant populations. VI. An entomophilous species, *Potentilla erecta*, in two contrasting habitats. *Heredity* **24**, 407–422.

Whitehead, D. R. 1983 Wind pollination: some ecological and evolutionary perspectives. In *Pollination biology* (ed. L. Real), pp. 97–108. Orlando, Florida: Academic Press.

Willemstein, S. C. 1987 *An evolutionary basis for pollination ecology*. Leiden University Press.

Willson, M. F. 1979 Sexual selection in plants. *Am. Nat.* **113**, 777–790.

Discussion

P. J. Grubb (*Botany School, University of Cambridge, U.K.*). None of Professor Cox's nine conditions favouring wind pollination over insect pollination seems to cover explicitly the preponderance of wind pollination in the trees of cool temperate deciduous forests, the grasses that dominate the steppes and the shrubs that dominate cold semi-deserts. Is it unreasonable to suggest that the major factor here has been the unreliability of insect populations in spring resulting from the effects of the winter cold?

P. A. Cox. Professor Grubb is correct in suggesting that unpredictable mortality of insect pollinators in cool temperate areas may favour wind-pollinated species. This idea extends my evolutionary arguments concerning pollinator extinction to an ecological arena of greater testability: possible higher variance in the reproductive success of entomophilous species as compared with anemophilous species due solely to fluctuations in pollinator populations.

Why do so few animals form endosymbiotic associations with photosynthetic microbes?

D. C. SMITH

University of Edinburgh, Old College, South Bridge, Edinburgh, EH8 9YL, U.K.

SUMMARY

A survey of modern associations in which protists or invertebrates are hosts shows that very few of the many species of photosynthetic microbes are adapted to an endosymbiotic existence. None occurs as intracellular symbionts in animals structurally more complex than cnidarians and platyhelminths. Photosynthetic symbionts are not usually capable of being the sole food source for hosts because they do not provide a balanced diet; most hosts therefore retain holozoic feeding. Interactions between hosts and intracellular symbionts are complex, and have to include mechanisms for inducing release of photosynthate from symbionts as well as controlling symbiont cell division. Possession of symbionts imposes a measurable cost on hosts. For the great majority of animals, the costs of adapting to herbivory or other forms of nutrition are probably less than that of hosting photosynthetic symbionts, especially when the need for exposure of a large surface area to light is borne in mind. Once hosts become multicellular, it is virtually impossible for any photosynthetic symbionts they possess to evolve into organelles because they are restricted to specific host cell types. After the evolution of the eukaryotic ancestors of plants in the late Precambrian, the major significance of endosymbiosis to the evolution of plant–animal interactions has been the development of gut symbioses in some major groups of herbivores.

1. INTRODUCTION

This paper deals with endosymbiotic associations between photosynthetic microbes and non-photosynthetic organisms. In particular, it addresses the question of why this type of symbiosis, so important in the early evolution of some major eukaryote groups in the Precambrian, has been of only very minor significance in subsequent evolution, contributing to the success of just a limited number of animals groups, mostly lower invertebrates and all aquatic.

The main types of modern associations will be surveyed, and the complexities of interactions between host and symbiont will be analysed. The term 'photosynthetic microbe' will include both cyanobacteria and unicellular algae; 'endosymbiosis' will refer to associations where the symbionts are within the body of the host, being either intracellular, extracellular between host cells, or contained in cavities. Some reference will be made to other types of symbiosis in order to reinforce the conclusions that are reached.

2. MODERN ASSOCIATIONS BETWEEN PHOTOSYNTHETIC MICROBES AND LOWER INVERTEBRATES OR PROTISTA
(a) Protists as hosts

Non-photosynthetic prokaryote symbionts occur quite widely (Lee *et al.* 1985), suggesting that symbiosis is a common strategy for protists. With regard to photosynthetic symbionts, in freshwater habitats at least 30 genera of ciliates as well as various amoebae have *Chlorella* symbionts (Reisser & Wiessner, 1984).

Ecologically, one of the most important groups is symbiotic benthic marine foraminifera which carpet extensive areas of certain shallow water habitats; they have a variety of algal symbionts, including dinoflagellates, chlorophytes and various diatoms. Radiolarians, whose symbionts are mostly dinoflagellates, are sometimes major contributors to the planktonic zone.

A variety of protists, distributed amongst at least seven genera, also form associations with cyanobacteria. The symbionts are unculturable, and are usually termed cyanelles. The cyanelles of *Cyanophora paradoxa* have a very low genome size (comparable to that of plastids), and their $16s$ rRNA sequence suggests a close alliance to plastids.

In all cases, the photosynthetic symbionts are intracellular, each symbiont cell being surrounded by a host membrane.

(b) Cnidarians

Cnidarians have only algal and not cyanobacterial symbionts, and these are normally intracellular within the digestive cells of the gastrodermis, each symbiont being surrounded by a host membrane. In a few sea anemones, symbionts may also occur in the mesoglea. The only freshwater cnidarians are *Hydra*, some species of which have *Chlorella* symbionts and are collectively known as green hydra.

Ecologically, by far the most important group are the hermatypic corals, mostly comprised of *Scleractinia*, which all have symbionts belonging to the genus *Symbiodinium*. Coral reefs cover about 2×10^8 km² of the earth's surface. The presence of symbionts is responsible

Phil. Trans. R. Soc. Lond. B (1991) **333**, 225–230
Printed in Great Britain

both for the high primary productivity of reefs (about 10 to 50 times greater than that of the surrounding tropical waters), and for enhancing the rate of calcification or corals to high levels, enabling them to exploit their habitats successfully (Goreau 1961).

(c) Porifera

Eight of the 18 genera of freshwater sponges which have been examined possess *Chlorella* symbionts. Symbiotic sponges are more prominent in shallow-water marine habitats, where many species contain an abundant population of both intracellular and extracellular prokaryotic symbionts, including cyanobacteria as well as bacteria.

(d) Molluscs

Few genera of molluscs possess photosynthetic symbionts. Ecologically the most prominent are giant clams, which can occur abundantly on indo-pacific reefs, and which all possess *Symbiodinium* symbionts located extracellularly within the blood vessels of the mantles. The two freshwater genera that are symbiotic, *Anodonta* and *Unio*, have *Chlorella* symbionts, also extracellular within the body of the host.

Some nudibranchs such as *Elysia viridis* sequester chloroplasts from the seaweeds (usually members of the Siphonales) upon which they feed. These chloroplasts continue photosynthesis for prolonged periods after entering host digestive cells (Hinde & Smith 1975) and this phenomenon has been termed 'chloroplast symbiosis'. In an analogous way, some aeolian opisthobranch molluscs acquire *Symbiodinium* from feeding on symbiotic alcyonaria (Rudman 1981).

(e) Platyhelminths

Examples of symbiosis are scarce. A few freshwater neorhabdocoel turbellarians have *Chlorella* endosymbionts. Amongst marine forms, only the acoel turbellarian *Convoluta roscoffensis* is of sufficient ecological prominence to occur abundantly in certain habitats. Its endosymbiont, which is intracellular, is *Tetraselmis*.

(f) Ascidians

Some marine ascidians have *Prochloron* (now considered to belong to the cyanobacteria) as extracellular symbionts, but these are rarely so common as to dominate habitats. *Prochloron* occurs in the cloaca and test in members of at least five genera of the family Didemnidae, and also on the surface of certain members of this family and three others (Polyclinidae, Polycitoridae and Styelidae). Some ascidians have a variety of other cyanobacteria, particularly *Synechocystis* on their surface as well as prochlorons.

3. CONCLUSIONS FROM MODERN ASSOCIATIONS

The following conclusions can be drawn from the survey of modern associations.

(a) Few photosynthetic microbes successfully adapt to endosymbiosis

There are very many species of photosynthetic microbes in aquatic habitats, but very few live endosymbiotically in animals or protists. The only eukaryotic unicellular algae that exist as intracellular or extracellular endosymbionts in a range of different types of freshwater hosts can all be assigned to the *Chlorella vulgaris* group (comprising *C. vulgaris*, *C. sorokiniana* and *C. saccarophila*) (Douglas & Huss 1986). In marine habitats, the range is a little wider, extending to a number of *Symbiodinium* spp. (Trench & Blank 1987) and one or two other dinoflagellate genera. Apart from the cyanelles of protists and the symbionts of some marine sponges, cyanobacteria do not occur as intracellular symbionts in animals.

The fact that very few photosynthetic microbes are adapted to an intracellular existence agrees closely with observations on symbiosis in general, where the range of endosymbionts which are extracellular is greater than those which are intracellular (Smith & Douglas 1987).

In marine habitats where several different types of photosynthetic symbiont occur, it is nevertheless rare for host cells – and indeed host animals – permanently to accommodate several different types of photosynthetic symbiont at the same time. The only exceptions are foraminifera in general, and the flatworm *Haplodiscus* (Trench & Winsor 1987) in whose cells the two dinoflagellates *Symbiodinium* and *Amphidinium* can coexist. Muscatine (1971) reported the presence of both *Symbiodinium* and an unidentified green algal symbiont in some specimens of sea anemones of the genus *Anthopleura* (although he speculated that the green algae might be a 'parasitic' infection).

In the case of foraminifera, Lee (1974) and Lee *et al.* (1979) speculate that the simultaneous presence of different symbiont types within a single host may have evolved from their unusual habit of gathering food and storing it for later digestion, so that digestion-resistant organisms would have a greater chance of survival in the host.

The extreme paucity of photosynthetic microbes capable of a permanent intracellular symbiotic existence agrees with the belief that despite the wide variety of algal and plant groups, modern plastids are derived from very few ancestral lines.

(b) Photosynthetic endosymbioses are restricted to aquatic habitats

The symbioses only become ecologically important (i.e. as dominant organisms in significant communities) in marine habitats, especially tropical. Douglas (1991), reviewing the evolutionary significance of endosymbioses, points out that hosts may be able to extend their ecological range by acquiring endosymbionts with novel metabolic capabilities. Thus, the success of hermatypic corals is due to the acquisition through their symbionts both of a supply of photosynthetically fixed carbon and of processes which lead to a very substantial enhancement of calcium carbonate deposition.

(c) *Intracellular photosynthetic symbionts are restricted to relatively few animal groups*

Intracellular photosynthetic symbionts are only at all common in certain groups of protists and cnidarians. Some of the cyanobacterial symbionts of sponges are intracellular, as are probably the few algal symbionts of platyhelminths which are symbiotic. Even when non-photosynthetic microbes are considered, they are commonly intracellular only in the lower animal groups with the exception of arthropods – especially mycetocyte-bearing insects. The general absence of intracellular symbionts from most other evolutionarily advanced groups may well reflect that they have reached a stage of biochemical and morphological differentiation where little or no net selective advantage would be gained from the possession of intracellular symbionts.

(d) *Photosynthetic endosymbionts usually supplement rather than replace holozoic feeding*

Hardly any hosts lose the capacity for holozoic feeding, so that photosynthetic symbionts usually supplement rather than replace conventional sources of food. This is not surprising as photosynthate is released from symbionts to hosts as only one or a few types of molecule (see below), so providing an incomplete diet and therefore being unsatisfactory as a sole food source. For example, *Chlorella* symbionts release only maltose. Although this sugar might seem little better than 'junk food', it enables the host to divert amino acids from respiration and gluconeogenesis to growth processes, and this can be particularly valuable in habitats where nitrogen sources are scarce (Rees & Ellard 1989).

The few associations in which adult hosts cease feeding holozoically include adult *Convoluta roscoffensis* which lives in nutrient-rich habitats, *Aiptasia* and xeniid corals. However, in the symbioses between chemoautotrophic bacteria and marine organisms such as the pogonophorans of deep-sea hydrothermal vents, the digestive systems of most types of adult host become non-functional, although juveniles feed.

4. INTERACTIONS BETWEEN INTRACELLULAR SYMBIONTS AND HOST CELLS ARE COMPLEX

For a photosynthetic symbiont to exist permanently inside a host cell, a variety of complex interactions must occur between the partners, and this may help to explain why so few photosynthetic microbes have become successful symbionts. These interactions may be summarized as follows.

(a) *Living cells of the symbiont are induced to release photosynthate in substantial quantities*

Host cells do not normally acquire organic nutrients from intracellular symbionts by digesting them. Hence,

mechanisms must develop whereby symbionts release photosynthate to a much greater extent than free-living algae; release in symbiosis may be substantial, and in some corals may constitute over 90% of total carbon fixed in photosynthesis (Muscatine *et al.* 1984). Although photosynthate release from symbionts has been one of the most studied aspects of symbiosis, the phenomenon is still not fully understood. Factors inducing release have been partly identified, but the mechanisms remain obscure. Furthermore, these factors differ with different types of symbiont. Cells of isolated symbiotic *Chlorella* release 50–60% of their photosynthetically fixed carbon when maintained at pH 4.0; but at pH 7.0, release is negligible (Douglas & Smith 1984). By contrast, release from isolated *Symbiodinium* symbionts varies little with pH but is markedly stimulated by thermolabile, water-soluble 'factors' present in homogenates of host tissue (see, for example, Sutton & Hoegh-Guldberg (1990)); 'factors' from one type of host can induce release from the symbionts of another type (Muscatine, 1967). These 'factors', which have not been characterized, have no effect on free-living algae, nor on *Chlorella* nor even on the photosynthetic symbionts of lichens; in addition, 'factors' cannot be isolated from non-symbiotic hosts. In lichens, the fact that symbionts are extracellular should make them easier to study experimentally, but even so, the process remains poorly understood and release appears to depend on some aspect of physical contact between symbiont and host.

For all types of photosynthetic symbiont, only one or a very few products of photosynthesis are released to the host cell: maltose from *Chlorella* symbionts; glycerol, alanine and fatty acids from *Symbiodinium*; glucose from 'symbiotic chloroplasts'; and either glucose or a polyol from symbionts of lichens.

(b) *Host cell and intracellular symbiont have to compete with each other for scarce essential nutrients*

Some of the key nutrients essential for the growth of host and symbiont are often in restricted supply in aquatic habitats; indeed, nutrient shortage is often regarded as encouraging colonization by symbiotic associations. At the cell level, mechanisms therefore have to develop so that competition for scarce nutrients such as phosphorus is sufficiently balanced to enable both partners to continue growing. Thus, when green hydra is starved, both hosts and symbiont stop growing (Ellard 1986). The need to acquire scarce nutrients may also help explain why so few animal hosts forsake holozoic feeding mechanisms. The value of the symbionts to the host in situations where nitrogen is limiting is that, as noted above, provision of ample fixed carbon through symbiont photosynthesis 'spares' amino acids from having to be used as substrates for respiration or for gluconeogenesis (Rees & Ellard 1989).

(c) *Cell division of symbionts must be controlled*

The growth rate of photosynthetic symbionts within host cells is usually slower than growth rates in culture by one or more factors of ten (Smith & Douglas 1987). For example, symbiotic *Chlorella* increase at a rate of 2.3 cells cell^{-1} day^{-1} in culture, but only 0.1 cells cell^{-1} day^{-1} in symbiosis with green hydra (Douglas & Huss 1986). A variety of hypothetical explanations have been offered on the basis of experimental observations. They include: restriction of nutrient supply to symbionts by host cell (Rees 1986); restriction in the supply of some 'factor' essential to symbiont cell division (McAuley 1985); diversion of fixed carbon from growth to translocation to host (Douglas & Smith 1984); and variation in factors known to affect symbiont cell-division such as pH in the case of *Chlorella* symbionts (Douglas & Smith 1984).

(d) Size and location of the symbiont population must be regulated

As well as controlling symbiont cell-division, mechanisms are needed to regulate both the overall size of the symbiont population and its location (not only to specific host cell types or morphological regions, but also specific regions of the host cell). Again, such mechanisms are not understood.

(e) Mechanisms are required for symbionts to be transmitted to succeeding host generations

In the case of protist hosts, transmission of intracellular symbionts to succeeding generations routinely occurs during host cell division. Allocation of symbionts to host daughter cells may be precise when the number per cell is small (for example, as in *Paulinella*), or more randomly when symbionts are numerous and the probability of transfer to a daughter cell very high (as in, for example, *Paramecium*).

In metazoan hosts with sexual reproduction, mechanisms are required which ensure either that symbionts are transmitted with the egg, or that they are acquired afresh from the environment at each generation. In the latter case, it must be remembered that few if any of the common intracellular photosynthetic symbionts can exist permanently and abundantly independent of the host (perhaps not surprising, because the intracellular environment to which they are so successfully adapted is presumably very different from that external to the host).

In green hydra, symbionts may become exocytosed from gastrodermal cells into the mesoglea and then phagocytosed by the epidermis-derived oocytes (Campbell 1990). In corals, some species rely on acquisition from the environment whereas others (which may even be from the same genus) can transmit symbionts directly to eggs (see review by Fitt (1984)). On coral reefs, most hosts release *Symbiodinium* symbionts into the surrounding seawater; possibly either as a mechanism for regulating symbiont population size, or providing a supply of symbionts for those animals who have to acquire symbionts from the environment each sexual generation. Giant clams rely entirely on acquisition from the environment: because of their growth habit, embedded in crevices on reefs, it would be a considerable advantage if they could acquire symbionts released from corals, implying that symbionts could be adapted to existence in a variety of different host types. In general, specificity of some strains of *Symbiodinium* symbionts for their hosts seems relatively low (Trench & Schoenberg 1976; Trench *et al.* 1981).

In *Convoluta roscoffensis*, symbionts are deposited onto the surface of the egg capsule, and so are available for young larvae when they hatch.

(f) Hosts must afford the cost of sustaining their symbionts

The variety of interactions listed previously impose an undoubted cost upon hosts, which can be shown by comparing the growth of hosts in the dark with and without symbionts (see, for example, Douglas & Smith (1983) for green hydra). The cost of bearing symbionts has to be set against the benefits they bring to the host. For example, nitrogen-fixing organelles probably never evolved because of the high energy costs of nitrogen fixation (and would have been particularly high for an ancestral host which was a unicell and therefore already had a relatively high specific metabolic rate), coupled with the fact that the nitrogenase enzyme is inactivated by exposure to molecular oxygen.

The cost of maintaining photosynthetic endosymbionts may be greater for most animals than the costs of adapting to the digestion of plants or the capture of herbivores. This may be a reason why so few have adopted this symbiosis as an evolutionary strategy. Indeed, the animal group in which photosynthetic symbionts are most widespread, the cnidarians, already have a plant-like sessile habit, often exposing relatively large surface areas because they are also filter feeders.

5. THE REQUIREMENT FOR EXPOSURE OF PHOTOSYNTHETIC SYMBIONTS TO LIGHT

Dependence upon photosynthetic symbionts carries with it requirement for prolonged periods of exposure to light of a relatively large surface area of host. For unicellular or cnidarian hosts, which naturally have large surface area to volume ratios, this presents few problems. Structurally more complex organisms have to develop adaptations both exposing large surface areas, and to reducing the risk this brings of attack by predators. Thus, a variety of adaptations can be seen in molluscs. The symbiont-containing mantle of giant clams is extremely sensitive to shading, and is withdrawn inside the shell within seconds of a sudden reduction in light intensity. The freshwater bivalve *Anodonta* makes use of the fact that damage to the shell causes transparent areas to develop, which then allow light through to its *Chlorella* symbionts while still retaining the defensive advantage of a shell. The nudibranch *Elysia viridis*, which acquires 'symbiotic chloroplasts' from its food plant, is believed to benefit from the camouflage of its consequential deep green colour when it is feeding on its host seaweed. Its tropical relative, *Placobranchus ianthobapsus* can either expose its plastid containing tissues, or fold them up so they are not visible (Trench 1975 and personal communication).

6. EVOLUTION OF PHOTOSYNTHETIC SYMBIONTS INTO ORGANELLES IS PROBABLY IMPOSSIBLE IN MULTICELLULAR ORGANISMS

Douglas (1991) states that a symbiont can be considered to have evolved into an organelle if genes essential to its functioning are transferred to the host nucleus. The host nuclei most likely initially to acquire such DNA are those of the cells containing the symbionts. However, the transferred sequence of symbiont DNA will only persist in succeeding generations if the recipient nuclei also contribute to the reproductive propagules of the host. Douglas points out this is the norm in unicellular protists, but virtually unknown in sexually reproducing multicellular hosts. Hence, once multicellular organisms evolved, it became most unlikely that any further new organelles could evolve from symbionts within them.

Incidentally, Douglas' statement clarifies the confusing situation concerning the extent to which the cyanelles of certain protists can be considered organelles. In *Cyanophora paradoxa*, the cyanelles have a genome size of 127 000 base pairs, comparable to that of plastids (Wassman *et al.* 1987), and this has led to questions of the extent to which they differ from plastids. However, gene transfer from cyanelles to host nucleus has not been shown; a key feature is that the small subunits of ribulose bisphosphate carboxylase are coded in the nucleus of chlorophytes and plants, but are coded in the cyanelles of *C. paradoxa* (Heinhorst & Shively 1983).

7. DISCUSSION

There are various reasons why symbiosis between animals and photosynthetic microbes is not widespread. Very few of the many species of photosynthetic unicells can exist as successful intracellular symbionts, and the choice of those that can associate with a range of organisms is virtually restricted to *Chlorella* in freshwater habitats, and to a small number in marine habitats, especially the dinoflagellates *Symbiodinium* and *Amphidinium*. Further, reliance on photosynthetically fixed carbon from symbionts as a food source means that hosts have to expose a relatively large surface area to light for long periods.

Interactions between intracellular photosynthetic symbionts and their hosts are complex. They impose costs upon the hosts, including those of mechanisms for control of symbiont cell division, and regulation of symbiont population size and location. However, once organisms became multicellular, it was more or less impossible for any symbionts they possessed to evolve into organelles unless symbionts were present in all host cells. Photosynthetic symbionts are invariably restricted to specific cell types in multicellular hosts.

Symbiosis with photosynthetic microbes has therefore been of very limited importance in animal evolution. However, quite another type of symbiosis, involving non-photosynthetic microbes, has been of major significance to evolutionary interactions between plants and animals. These are the associations with cellulose-degrading endosymbionts which occur in the guts of certain major groups of herbivores. Much of the fixed carbon which accumulates in plants is in the form of structural polysaccharides, especially cellulose. Successful vertebrate and some insect herbivores do not themselves produce cellulases, and depend to a greater or lesser extent on populations of digestive tract symbionts to break down the refractory plant material that comprises the bulk of their food.

Several authors (Bauchop 1977; McBee 1977; Douglas 1991) have noted that cellulolytic gut symbionts are more advantageous to large rather than small animals. Cellulose is degraded slowly, and the host has to allocate considerable space to the fermentation chamber containing symbionts and decomposing plant material (Alexander, this symposium). Small mammals such as rabbits and voles must pass food through the gut more rapidly and select more digestible food than large herbivores. For relatively very small herbivores such as insects, the cost of cellulolytic symbionts is much greater than for vertebrates. Hence, many insect herbivores produce their own cellulases, whereas others develop alternative strategies for exploiting microbial cellulases. For example, an ecologically important strategy is the degradation of plant material externally to insects, as in 'fungus gardening' by ambrosia beetles and by Macrotermitidae. The latter, together with various insects such as siricid wasps, scolytid and cerambycid beetles, not only feed on the fungi and bacteria which have degraded the cellulose, but also utilize free microbial cellulases in their guts. Such microbial cellulases are stable and contribute to the degradation of dietary cellulose for extended periods (Martin 1987, and this symposium).

Thus, symbiosis has had important evolutionary consequences in both the Precambrian and Phanerozoic eras, but the symbiotic acquisition of the capacity for photosynthesis ceased to be an important strategy after the rise of photosynthetic eukaryotes in the Precambrian era. Thereafter, herbivory, whether or not involving gut symbioses, was the major route for gaining access to photosynthetically fixed carbon.

I thank Dr Angela Douglas for many helpful and constructive comments.

REFERENCES

Bauchop, T. 1977 Foregut fermentation. In *Microbial ecology of the gut* (ed. R. T. J. Clarke & T. Bauchop), pp. 223–251. London: Academic Press.

Campbell, R. D. 1990 Transmission of symbiotic algae through sexual reproduction in hydra: movement of algae into the oocyte. *Tiss. Cell* **22**, 137–147.

Douglas, A. E. 1991 Symbiosis in Evolution. In *Oxf. Surv. Evol. Biol., Vol. 8* (ed. D. J. Futuyma & J. Antonovics). Oxford University Press. (In the press.)

Douglas, A. E. & Huss, V. A. R. 1986 On the characteristics and taxonomic position of symbiotic *Chlorella*. *Arch. Microbiol.* **145**, 80–84.

Douglas, A. E. & Smith, D. C. 1983 The cost of symbionts to the host in the green hydra symbiosis. In *Endocytobiology, endosymbiosis and cell biology II* (ed. W. Schwemmler &

H. E. A. Schenk), pp. 631–648. Berlin: Walter de Gruyter and Co.

Douglas, A. E. & Smith, D. C. 1984 The green hydra symbiosis VIII. Mechanisms in symbiont regulation. *Proc. R. Soc. Lond.* B **221**, 291–319.

Ellard, F. 1986 *Growth in symbiotic hydra.* D.Phil. thesis, University of Oxford.

Fitt, W. K. 1984 The role of chemosensory behaviour of *Symbiodinium microadriaticum*, intermediate hosts, and host behaviour in the infection of coelenterates and molluscs with zooxanthellae. *Mar. Biol.* **81**, 9–17.

Goreau, T. F. 1961 Problems of growth and calcium deposition in corals. *Endeavour* **20**, 32–39.

Heinhorst, S. & Shively, J. M. 1983 Encoding of both subunits of ribulose-1, 5-bisphosphate carboxylase by organelle genome of *Cyanophora paradoxa*. *Nature, Lond.* **304**, 373–374.

Hinde, R. & Smith, D. C. 1975 The role of photosynthesis in the nutrition of the mollusc *Elysia viridis*. *Biol. J. Linn. Soc.* **7**, 161–171.

Lee, J. J. 1974 Towards understanding the niche of Foraminifera. In *Foraminifera, Vol. 1* (ed. R. H. Hedley & G. Adams), pp. 208–257. New York: Academic Press.

Lee, J. J., McEnery, M. E., Kahn, E. G. & Schuster, F. L. 1979 Symbiosis and the evolution of the larger foraminifera. *Micropalaeontology* **25**, 118–140.

Lee, J. J., Soldo, A. T., Reisser, W., Lee, M. J., Jeon, K. W., & Gortz, H. D. 1985 The extent of algal and bacterial symbioses in Protozoa. *J. Protozool.* **32**, 391–403.

McAuley, P. J. 1985 The cell cycle of symbiotic *Chlorella*. I. The relationship between host feeding and algal cell growth and division. *J. Cell Sci.* **77**, 225–239.

McBee, R. H. 1977 Fermentation in the hindgut. In *Microbial ecology of the gut* (ed. R. T. J. Clarke & T. Bauchop), pp. 185–222. London: Academic Press.

Martin, M. M. 1987 *Invertebrate-microbial interactions.* Ithaca: Cornell University Press.

Muscatine, L. 1967 Glycerol excretion by symbiotic algae from corals and *Tridacna*. *Science, Wash.* **156**, 156–519.

Muscatine, L. 1971 Experiments on green algae coexistent with zooxanthellae in sea anemones. *Pacific Sci.* **XXV**, 13–21.

Muscatine, L., Falkowski, P., Porter, J. & Dubinsky, Z. 1984 Fate of photosynthetically fixed carbon in light and shade-adapted colonies of the symbiotic coral *Stylophora pistillata*. *Proc. R. Soc. Lond.* B **222**, 181–202.

Rees, T. A. V. 1986 The green hydra symbiosis and ammonium I. The role of the host in ammonium assimilation and its possible regulatory significance. *Proc. R. Soc. Lond.* B **229**, 299–314.

Rees, T. A. V. & Ellard, F. 1989 Nitrogen conservation and the green hydra symbiosis. *Proc. R. Soc. Lond.* **236**, 203–212.

Reisser, W. & Wiessner, W. 1984 Autotrophic freshwater symbionts. *Encyclop. Pl. Physiol., New Series* **17**, 59–74.

Rudman, W. D. 1981 The anatomy and biology of alcyonarian-feeding aeolid opisthobranch molluscs and their development of symbiosis with zooxanthellae. *Zool. J. Linn. Soc.* **72**, 219–262.

Smith, D. C. & Douglas, A. E. 1987 *The biology of symbiosis.* London: Edward Arnold.

Sutton, D. C. & Hoegh-Guldberg, O. 1990 Host-zooxanthella interactions in four temperate marine invertebrate symbioses: assessment of effect of host extract on symbionts. *Biol. Bull.* **178**, 175–186.

Trench, R. K. 1975 Of 'leaves that crawl': functional chloroplasts in animal cells. *Symp. Soc. exp. Biol.* **29**, 229–265.

Trench, R. K. & Blank, R. J. 1987 *Symbiodinium microadriaticum* Freudenthal, *S. goreauii* sp. nov., *S. kawagutii* sp. nov. and *S. pilosum* sp. nov.: gymnodinoid dinoflagellate symbionts of marine invertebrates. *J. Phycol.* **23**, 469–481.

Trench, R. K., Colley, N. J. & Fitt, W. K. 1981 Recognition phenomena in symbioses between marine invertebrates and 'zooxanthellae'; uptake, sequestration and persistence. *Ber. dt. bot. Ges.* **94**, 529–549.

Trench, R. K. & Schoenberg, D. A. 1976 Specificity of symbioses between marine cnidarians and zooxanthellae. In *Coelenterate ecology and behaviour* (ed. G. O. Mackie), pp. 423–432. New York: Plenum.

Wassman, C. C., Loffelhardt, W. & Bohnert, H. J. 1987 Cyanelles: organisation and molecular biology. In *The cyanobacteria – a comprehensive review* (ed. P. Fay & C. van Baalen), pp. 303–324. Amsterdam: Elsevier.

Discussion

E. A. BERNAYS (*Department of Entomology, University of Arizona, U.S.A.*). What is special about *Chlorella* that enables it to form most of the symbiotic interactions?

D. C. SMITH. A very recent paper by Kessler *et al.* (1991) examined the excretion of sugar (maltose, glucose or glucose-6-phosphate) by cultures of 38 strains of *Chlorella* belonging to 15 species of which seven are capable and eight incapable of symbiosis with green hydra. All excreted some sugar, but only in trace amounts by 24 strains, and relatively small amounts by a further eight strains. However, one non-symbiotic strain (*C. mirabilis*) excreted large amounts of maltose below pH 4, and several non-symbiotic strains of *C. vulgaris* had maximum rates of maltose excretion below pH 4. Kessler *et al.* therefore concluded that there was no obvious correlation between sugar excretion and the ability or inability of the *Chlorella* species to form stable symbioses with green hydra. All the species that can form a stable symbiosis are those that are capable of growing in acid media at or below pH 4.0. Hence, acid tolerance seems likely to be an important factor.

Reference

Kessler, E., Kauer, G. & M. Rahat 1991 Excretion of sugars by *Chlorella* species capable and incapable of symbiosis with *Hydra viridis*. *Bot. Acta*, **104**, 58–63.

Marine plants and their herbivores: coevolutionary myth and precarious mutualisms

R. N. HUGHES AND C. J. GLIDDON

School of Biological Sciences, University College of North Wales, Bangor, Gwynedd LL57 2UW, U.K.

SUMMARY

Planktonic and benthic algal–herbivore interactions are reviewed. In many cases, generation times of plant and herbivore are similar, yet there is no clear example of a coevolutionary arms race. Mutualisms have evolved, but these are vulnerable to extinction. We suggest that this vulnerability is caused by the loss of evolutionary flexibility accompanying the tightly linked, coadapted gene complexes.

1. INTRODUCTION

A study of interactions among species is the classical domain of the community ecologist. In particular, the elaboration by theoretical ecologists of the logistic equation has provided an interesting insight into the dynamics of competition, exploitation (e.g. plant–herbivore, prey–predator and host–parasite interactions) and mutualism. However, when attempting to relate empirical observations from nature to predictions of ecological theory, it is important to bear in mind the evolutionary context in which the species or populations of study are contained. In this sense, the species interactions can be viewed as examples of coevolution (*sensu* Ehrlich & Raven 1964). This early definition of coevolution embraced all evolution resulting from biological interactions among species, but is now more commonly called diffuse coevolution (Futuyma & Slatkin 1983). Coevolution is more frequently reserved for stepwise, reciprocal evolution (Janzen 1980) resulting in species tracking each other's response over long periods of geological time.

(a) *Some considerations from theory*

The majority of ecological models of species interactions have underlying assumptions, often unstated, which can make it hazardous to extend them to include evolution. For example, continuous-time models assume that the interacting species 'perceive' time in the same way, even though their evolutionary generation times may be very different. This problem can be removed by defining r in terms of the rate of increase per unit of absolute time. The problem re-emerges, however, if the evolution of the parameter r is to be considered, as this will occur with an evolutionary timebase perhaps more related to generation time than to absolute time. Discrete-time models clearly suffer from similar problems, as the finite difference equation for each species must relate to its generation time in evolutionary models.

In general, if we consider systems of differential equations describing interacting species, unless the individual time constants are very similar, it is possible to decouple the equations and treat them as though they were independent. If there is to be some emergent property of the species interactions, it is necessary that the ratio of the evolutionary generation times of the species should approach unity (for general models of species interactions, this ratio can be defined as the number of genotypes of species A encountered by a single genotype of species B). This requirement is approximately met among a wide range of marine plants and their herbivores.

(b) *Marine–terrestrial comparisons*

Marine benthic and terrestrial communities are characterized by interactions among dissimilar organisms, with independent evolutionary histories involving different phylogenetic and environmental constraints. Both, however, sustain broadly similar biological interactions, which have been the subject of several incisive reviews (Hay *et al.* 1987; Hay & Fenical 1988; Pfister & Hay 1988). Most marine benthic herbivores have broad diets, whereas most terrestrial herbivores are relatively specialized. This difference may reflect the contrasted mechanisms of recruitment in the two environments. Insects constitute the majority of grazers among terrestrial communities. The adults disperse and oviposit on precisely selected host plants, which may be relatively rare and inconspicuous. Marine macroherbivores, mostly fishes, sea urchins and large molluscs, liberate planktonic larvae, dispersed passively in water currents and unable to respond to locational stimuli over anything but the shortest distances (Butman 1987). Mesoherbivores, including small crustaceans, polychaetes and snails, disperse as larvae or as adults, but in either case there is usually a large component of undirected drift. Hence, whereas insects can be sure of locating a precise target and afford to be specialists, marine grazers generally cannot (Hay & Fenical 1988).

Macroherbivores have generation times similar to, or greater than most macroalgae upon which they feed and in mesoherbivores generation times are similar to, or less than those of their host algae (Hay & Fenical 1988). Generations times of marine benthic herbivores and macroalgae, therefore, are more closely matched than those of insects and flowering plants, and so might be expected to be more conducive to coevolution. For the same reason, this may also be expected of planktonic herbivores and microalgae. We shall show, however, that there is no more evidence of coevolution between herbivores and plants in aquatic than in terrestrial environments and we discuss the restrictive conditions likely to promote reciprocal evolution.

2. THE COEVOLUTIONARY MYTH
(a) *Herbivory and antiherbivory*

Macroalgae are important members of intertidal and shallow, subtidal, rocky habitats throughout the world, where they may contribute substantially to primary production. Much of this enters the detritus food chain (Mann 1973) and, in some situations, relatively little appears to be taken by herbivores. For example, on British coasts, dense zones of fucoids and kelp often show only sporadic signs of being heavily grazed. In contrast, tropical hard substrata are usually sparsely vegetated, but when herbivorous fish are excluded, a richer algal canopy develops (Lewis 1986). Such geographical variations suggest that macroalgae are well defended against all but the most powerful herbivores, mainly certain fishes and sea urchins. Herbivorous fishes are largely confined to warmer seas (Horn 1989), whereas sea urchins extend also into cold-temperature regions where they may exert strong grazing pressure subtidally (Vadas 1990). A few specialized gastropods also damage macroalgae (Hughes 1986), but snails are usually relatively mild grazers. Increasing evidence, however, reveals that small mesoherbivores, including isopods, amphipods, crabs, polychaetes and snails, consume far more macroalgal tissue than previously suspected (Brawley 1991). Clearly, grazers pose a general threat to macroalgae, which show a variety of defences. Among these defences, secondary metabolites are of renowned importance.

Like terrestrial vegetation, seaweeds produce a spectrum of secondary metabolites including terpenes, aromatic compounds and polyphenolics. Unlike terrestrial plants, seaweeds often incorporate halogens into these compounds and, being nitrogen limited, do not produce alkaloids (Hay & Fenical 1988). Rhodophytes widely produce aliphatic halo-ketones, brominated phenols and terpenes. Chlorophytes are typified by ephemeral, undefended species, but the order Caulerpales contains perennial species producing terpenoids. Phaeophytes produce polyphenolics, differing from those in terrestrial plants in being derived from phloroglucinol, hence the term 'phlorotannins' used

for these compounds. Whereas the secondary metabolites of red and green algae are lipophilic and can be applied via organic solvents to the surface of otherwise palatable plants, the phlorotannins are water soluble and must be immobilized in agar for presentation to herbivores. Feeding trials using these techniques have abundantly shown that many secondary metabolites deter herbivores (Hay & Fenical 1988).

However, despite burgeoning data showing that particular secondary metabolites repel grazers (Hay & Fenical 1988), repellency is not predictable from molecular structure. Compounds that deter one species may have no effect on another (Steinberg 1988) and seemingly trivial changes in molecular arrangement may prove to be critical. Dictyol-B and dictyol-E differ only in the position of one hydroxyl group, yet the former deters grazing by the rabbitfish *Siganus doliatus*, whereas the latter does not (Hay 1991; Hay *et al.* 1988 *a*). Repellent algae, however, mainly produce a range of secondary metabolites, resulting in a general effectiveness against macroherbivores. Overall, the picture is one of defence on a broad front, not of reciprocal, one-to-one combat between macroalgae and macroherbivores.

Less widespread in the phytoplankton than among benthic macroalgae, toxicity is a property of certain freshwater cyanobacteria and marine dinoflagellates. Planktonic cyanobacteria and algal protists have generation times in the order of 0.08–4.1 days (Fogg & Thake 1987). They are eaten by zooplanktonic cohabitants, including ciliates, rotifers, crustaceans and pelagic tunicates, which have generation times of some 0.5–30 days (Fenchel 1974; Allan 1976). Because the ranges of generation time are displaced only by an order of magnitude, the zooplanktonic populations closely track the phytoplanktonic and therefore are usually the principal source of plant mortality (Steele 1974). Not surprisingly, phytoplanktonic species often have attributes, including unmanageable dimensions, undigestibility and toxicity, that reduce vulnerability to the zooplankton (Porter 1977).

Toxicity can be effective against passively filtering cladocerans which temporarily cease feeding. The response, however, is variable; lake *Daphnia* respond quickly to *Microcystis*, whereas pond *Daphnia*, normally living in the absence of this toxic species, respond more slowly and consequently suffer greater mortality (DeMott *et al.* 1991).

Toxicity, however, is particularly appropriate against raptorial feeders, as these generally are more sensitive to food quality (Porter 1977). Raptorially feeding copepods grasp individual particles and although several items may accumulate in the mouth parts to be handled simultaneously (Vanderploeg *et al.* 1990), the individual nature of initial capture makes it feasible to select or reject items by their nutritional quality or toxicity. *Diaptomus birgei* rejects live *Microcystis aeruginosa* but consumes dead clumps that presumably have lost their toxicity (DeMott & Moxter 1991).

In the marine phytoplankton, toxicity is the hallmark of certain dinoflagellates. Unlike endotoxic freshwater cyanobacteria, these dinoflagellates secrete exotoxins, whose concentration in the seawater may reach dangerous proportions for copepods. *Calanus pacificus* grows poorly through its non-feeding, naupliar stages and suffers greater mortality when in the presence of toxic *Phytodiscus brevis* or *Gonyaulax grindleyi* (Huntley *et al.* 1987). In subsequent feeding stages *C. pacificus* discriminates against *G. grindleyi*, but a small proportion of cells may accidentally be ingested. Regurgitation, after a lag of some 45–120 min, minimizes the impact of the toxins (Sykes & Huntley 1987).

The effect of dinoflagellate toxins on marine copepods is similar to that of cyanobacterial toxins on freshwater cladocerans. Feeding rates are depressed and the secondary effects of starvation, reduced growth rate and lowered fecundity alleviate grazing pressure at the population level. This is enforced more acutely when exotoxins directly reduce growth and survivorship of naupliar stages. Together, these factors add positive feedback when hydrographical conditions favour growth of the dinoflagellate populations and so may contribute to the formation of blooms (Huntley *et al.* 1986).

As with the chemical defences of macroalgae, phytoplanktonic toxins are unpredictable in their effects. One toxin may affect certain grazers and not others, and a given grazer may be inhibited by some toxins and not others (Gilbert 1990). Apart from the weak correlation between sensitivity of *Daphnia* to toxic cyanobacteria and incidence of these bacteria in the habitat (DeMott *et al.* 1991; Gilbert 1990), there is no evidence of any evolutionary counter-response of zooplankters to the threat of poisoning.

(b) *Sequestering defences*

Macroherbivores are usually highly mobile, covering relatively large distances over which they may encounter a wide variety of algae. Here, there is scope for selective feeding during each foraging bout, when unpalatable algae may be avoided. Mesoherbivores, by contrast, are far less mobile and often the food plant is also their home and habitat. If this plant is unpalatable to macroherbivores, especially fishes, the associated mesoherbivores will be less at risk to predation than they would on an unprotected plant. Accordingly, mesoherbivores tend to exploit algae that are repellent to macroherbivores, so gaining 'enemy-free space' (Price *et al.* 1980).

Unless they become exceptionally numerous, mesoherbivores do not seriously damage their host algae, being kept far below carrying capacity by predators (Nelson 1979; Stoner 1980; Carpenter 1986). It is not surprising, therefore, that algae fail to produce secondary metabolites effective against mesoherbivores. Dictyol-E and pachydictyol-A repel herbivorous fish and urchins, but not the amphipod *Amphithoe longimana* or the polychaete *Platynereis dumerilii*, both of which prefer chemically defended algae such as *Dictyota dichotoma* and *Sargassum filipendula* (Hay *et al.* 1988 *b*). Indeed, secondary metabolites repelling macroherbivores may become specifically attractive to mesoherbivores.

The Caribbean amphipod *Pseudamphithoides incurvaria* specializes on chemically defended algae such as *Dictyota bartayresii*. Not only does the amphipod use these algae for food and shelter, but also cuts out pieces to construct a portable case in which to live (Hay *et al.* 1990). *P. incurvaria* is reluctant to use algae, even species of *Dictyota*, lacking diterpenes that repel herbivorous fish. Moreover, when presented with the normally palatable chlorophyte, *Ulva* spp., treated with differing concentrations of pachydictyol-A, *P. incurvaria* responds more readily to higher dosages, but when forced to build domiciles from untreated *Ulva*, the amphipod becomes vulnerable to fish predation.

Metabolic sequestering of secondary compounds occurs in a number of opisthobranchs, including the sea hare, *Aplysia californica*, which accumulates toxins from its food plants, *Laurencia* spp. and *Plocamium* spp., so gaining protection from predators (Stallard & Faulkner 1974). The most specialized feeders of all, however, are the ascoglossans. These sea slugs lance plant cells and suck out the sap. Each species specializes on algae with a particular morphology. The radular tooth forms a precision tool, suited to the shape and size of algal cell to be pierced, and in those slugs feeding on filamentous algae, the foot, likewise is moulded to the dimensions of the filament (Jensen 1983).

Ascoglossans are predominantly green, owing to the chloroplasts they have ingested, sequestered and maintained in diverticulae of the digestive gland (Clark & Busacca 1978). Photosynthates of the chloroplasts are an important resource for the slugs, whose dorsal flaps and frills increase the light-capturing surface area (Hughes 1986).

When feeding on chemically defended algae, ascoglossans sequester secondary metabolites, in addition to chloroplasts. *Elysia halimedae* selects the young, diterpenoid-rich, growing edges of its host plant, *Halimeda macroloba* (Paul & Van Alstyne 1988). When attacked, the slug secretes a repellent mucus, containing a sequestered and slightly modified diterpenoid that deters fish.

The sequestering of secondary metabolites seems not to be a round of coevolution in favour of the mesoherbivores, but the specialized exploitation of resources that happen to be a ubiquitous, permanent feature of the habitat. The resources themselves appeared as an evolutionary response of the host algae to a completely different set of grazers, the macroherbivores.

We conclude that among diverse taxa of aquatic herbivores and plants, offensive and defensive strategies have evolved not in reciprocal escalation, but as parallel responses to selection pressures associated with the general problems of eating (Horn 1989) and being eaten (Norton & Manley 1990). History and phylogeny will have played important roles in shaping these responses.

(c) *The fossil record*

Inferences about coevolution, based on present circumstances, usually are severely weakened by a lack of evidence from the past. Fortunately, there is an exception among marine plant–herbivore interactions. Calcareous, encrusting algae fossilize well and sometimes bear grazing marks clearly attributable to particular herbivores, also well represented as fossils. If there has been any stepwise, reciprocal evolution between these plants and their herbivores, it should be evident from the geological record (Steneck 1983, 1986, 1991).

Encrusting coralline algae, cosmopolitan occupants of hard substrata in shallow seas, are grazed by herbivores able to excavate the armoured thalli. Even such powerful grazers cannot normally inflict fatal damage and encrusting corallines typically predominate in areas of high grazing intensity. Armoured defence is achieved at the cost of slower growth (Steneck 1985) and greater susceptibility to fouling (Paine & Vadas 1969). Consequently, many encrusting corallines depend on grazers to prevent competitive exclusion and smothering (Steneck 1983).

In temperate regions, urchins and limpets graze the encrusting corallines. Limpets sometimes have become specifically associated with host plants (Estes & Steinberg 1988), probably because they use the host as habitat in addition to food. The wide-mouthed shell requires a close fit to the substratum if dislodgement by predators or water movement is to be resisted effectively (Hahn & Denny 1989). In the North Atlantic, *Tectura testudinalis* has a mutualistic relationship with the thick, encrusting coralline *Clathromorphum circumscriptum*, in which the limpet gains greater resistance to predators and the coralline is kept clear of fouling organisms (Steneck 1982). Are the defensive properties of corallines, the excavating ability of limpets and mutualism the result of stepwise, reciprocal evolution?

Encrusting algae resembling corallines originated in the Precambrian (Grant *et al.* 1991), before any herbivores had yet appeared. The first coralline to be morphologically differentiated, as in modern species, was *Archaeolithophyllum*. It possessed cellular fusion, a basal hypothallus, inner meristem and outer epithallus, and conceptacles enclosing spores and gametes. In the late Carboniferous, *Archaeolithophyllum* formed banks in moderately turbulent, shallow reef environments. The curly, discoidal thalli grew freely on sediments and were never cemented to hard substrata, as modern forms usually are. The suite of characters listed, enabled *Archaeolithophyllum* to recover from abrasion and breakage, caused by waves in stormy weather, but would become equally important among other corallines, millions of years later, in surviving damage inflicted by excavating herbivores.

Cellular fusion had evolved suddenly. It provided a mechanism for translocation, upon which differentiation into non-photosynthetic parts would depend, so paving the way for a radiation which even now may not yet have reached its limit. Key features in this radiation have been the potential for rapid lateral growth, an ability to encrust hard substrata, architectural diversity of the thallus, protection of spores and gametes within conceptacles and the capacity to regenerate tissues lost by deep wounding (Steneck 1983).

Despite possession of these characters by *Archaeo-lithophyllum* in the Palaeozoic, radiation of the encrusting corallines only gathered momentum some 100 million years later, towards the end of the Jurassic (Steneck 1983). Before the radiation, corallines had occupied sediments, and their fossils bear no grazing marks. Powerful, excavating herbivores were absent at that time and hard substrata were monopolized by fleshy algae, competitively superior to the corallines.

Limpets, urchins and herbivorous fishes, all powerful excavators, radiated in the Middle Mesozoic and early Cainozoic. They stripped the fleshy algae from hard substrata and tipped the competitive balance in favour of the encrusting corallines, whose own radiation ensued. 'True' limpets appeared in the late Triassic, preceding the radiation of encrusting corallines. They began to diversity in the Jurassic, but their grazing marks are found on the shells of bivalves and ammonites, where the limpets probably were feeding on epiphytes and endophytes. By the Cainozoic, limpet grazing marks also became relatively common on encrusting corallines (Steneck 1983).

The geological record, then, shows a sequence of independent evolutionary events. First, the evolution of translocation and morphological differentiation in encrusting corallines. This was selectively advantageous in turbulent, abrasive sedimentary habitats. Second, the evolution of deeply grazing limpets, urchins and fishes. This cleared fleshy algae from rocks and opened up a new habitat for the encrusting corallines, whose abrasion-resistant thalli and high regenerative capacity pre-adapted them to survive herbivory. Meanwhile, another major taxon of calcareous encrusting algae, the Solenoporaceae, lacking such pre-adaptations, became extinct (Steneck 1983). Neither the grazing mechanisms of the herbivores nor the principal resistive mechanisms of the coralline algae evolved in relation to each other.

During the extensive radiation of the encrusting corallines and coincidental radiation of herbivores, there was opportunity for some evolutionary 'fine tuning', in which the herbivores became more powerful, while the algae became morphologically less diverse and more resistant (Steneck 1986). Included in the fine tuning, were mutualisms such as that between *Clathromorphum circumscriptum* and *Tectura testudinalis*.

C. circumscriptum has an exceptionally thick, multi-layered epithallus that sustains grazing by *T. testudinalis*, yet protects the deeply underlying meristem. Throughout its benthic life, *T. testudinalis* survives better on the smooth, plane surface of *C. circumscriptum* than elsewhere, and *T. testudinalis* larvae probably are specifically attracted to the alga (Steneck 1982). *C. circumscriptum* usually is not found in the absence of *T. testudinalis*, because this limpet keeps its habitat and food supply free of epiphytes and staves off competitors. The thickened epithallus of the alga and host-specificity of the limpet seem to have evolved specifically, resulting in mutualism. Fossil evidence shows this to have happened relatively recently, in the Holocene (Steneck 1991).

In the case reviewed, there is no evidence of any arms race, but mutualism seems to have evolved as a result of historical happenstance.

3. THE EVOLUTIONARY TRANSIENCE AND FORTUITOUS ORIGINS OF MUTUALISMS

(a) *Limpets and host plants*

Other specific associations between limpets and host plants are geologically recent, probably having originated less than three million years ago (Estes & Steinberg 1988). They occur sporadically among taxa (Muñoz & Santelices 1989), suggesting that specific associations and mutualisms are evolutionary fleeting couplings that arise suddenly, only to disappear later when one or both partners become extinct. *Lottia alveus*, for example, suffered widespread extinction in the western North Atlantic when its host, the seagrass *Zostera marina*, was almost wiped out by a pathogen in the 1930s (Lindberg 1990; Steneck 1991).

(b) *Epiphytic cleaners*

Epiphytes, particularly encrusting invertebrates such as sponges, bryozoans and compound ascidians, sometimes occupy large areas of frondal surface (Seed & O'Connor 1981). This may interfere with photosynthesis (Oswald *et al.* 1984), decrease buoyancy (Wing & Clendenning 1971), and decrease flexibility (Dixon *et al.* 1981) or increase drag (Menge 1975), the last two increasing the risk of tearing or dislodgement. Epiphytic algae can have similar effects and when growing on encrusting forms, they may smother the host to death (Sousa 1979; Steneck 1982). By removing epiphytes, grazers improve the prospects of the host alga (Williams & Seed 1991). For example, *Rhodomela larix* survives, grows and reproduces better when its load of epiphytes is lessened by the grazing activities of amphipods and littorinid snails (D'Antonio 1985).

The change from an association that is only exploitative to one that is mutualistic, may involve but a modest evolutionary step. *Littorina obtusata* and *L. mariae* are sibling species, often occupying the same shores of the eastern North Atlantic, but always ecologically separated by exploiting different algae. *L. obtusata* exclusively uses the long-lived fucoid *Ascophyllum nodosum*. Juveniles feed on sporelings or other delicate epiphytes, but older, larger snails with a strong radular apparatus, graze the host's thallus (Norton & Manley 1990). Commensurate with the longevity of its host, *L. obtusata* is perennial. *L. mariae* on the other hand is an annual, growing only to a relatively small size, it is incapable of grazing the thalli of leathery macroalgae and retains a diet of small, delicate epiphytes and sporelings throughout its life. *L. mariae* lives on kelp laminae and particularly the fronds of *Fucus serratus*. Both substrata are ephemeral or annual, to which the short life cycle of *L. mariae* is suited. The dietary restriction to epiphytes, imposed by small body size, benefits *F. serratus*, however. Fronds grazed by *L. mariae* bear smaller epiphytic loads and consequently are less likely to be torn away by water movement (Williams 1990; Williams & Seed 1991). The mutualism between *L. mariae* and *F. serratus*, therefore, is simply the indirect result of interspecific competition between *L. mariae* and its sibling species, *L.*

obtusata, which has caused the evolution of niche separation through habitat choice and appropriate adjustment of life history.

(c) *Gardening*

Certain tropical-reef fishes, especially pomacentrids (Brawley & Adey 1977; Williams 1980) and temperate-zone limpets (Branch 1981) defend feeding territories, creating scope for controlled levels of grazing and the maximization of sustainable yield. Such 'gardening' is associated with a general scarcity of algal biomass and intense competition among grazers (Branch *et al.* 1991). Occupants rely entirely on their gardens for a supply of food. Consequently, gardening is limited to environments that do not interrupt primary production. For example, gardening limpets on the South African coast are restricted to low-shore levels, where algal production escapes seasonal limitation, and are more frequent on the relatively nutrient-poor east coast than on the enriched, upwelling region along the west coast. Similarly, gardening fishes predominate in the seasonally stable, but nutrient-depleted waters of tropical reefs, where there is intense herbivory (Vine 1974; Hatcher 1988).

Gardens are more productive than surrounding areas (Branch *et al.* 1991), partly because opportunistic, fast-growing algae are encouraged and partly because the algae are maintained in a rapid-growth phase by continual, but not excessive grazing. Horticultural-like practices are sometimes involved. Pomacentrids kill areas of coral, preparing suitable substratum for the settlement and germination of algal spores (Kaufman 1977; Wellington 1982) and some weed the garden by selectively tearing out inedible species (Lassuy 1980). Outside pomacentrid territories, the substratum is dominated by encrusting corallines and repellent foliose algae, whereas within the gardens grow delicate, filamentous forms such as *Polysiphonia* (Sammarco 1983). Unlike most other herbivorous reef fishes, pomacentrids are unable to triturate and digest physically or chemically defended algae, but with their closely spaced, needle-like teeth they are able to efficiently graze fine, filamentous forms. The garden therefore provides the fish with a continual supply of food and the algal crop with protection from overgrazing by other herbivores.

In contrast to pomacentrids, limpets increase above background the level of grazing within their territories (Branch *et al.* 1991). This relatively intense grazing, however, is not indiscriminate. *Patella longicosta* and *P. cochlear* crop the algae in such a way as to leave a reticulated pattern of ridges. This maximizes the surface area of plant growth and hence productivity (Branch 1981). The intense grazing favours fast growing, but competitively inferior species, *Ralfsia verrucosa* in the case of *P. longicosta*, *Herposiphonia heringii* and *Gelidium micropterum* in the case of *P. cochlear*. Outside territories, these algae are largely replaced by competitively dominant, encrusting forms, indeed *G. micropterum* has only been found within *P. cochlear* gardens (Branch 1975). Again, the garden is of mutual benefit to gardener and crop plant, the former gaining an assured supply of food, the latter a refuge from competitors.

Gardening results from contest competition for food and may disappear, even within a species, where competition is less intense. *Patella barbara* and *P. miniata* garden on the relatively impoverished east coast of South Africa, but graze non-territorially on the enriched west coast (Branch *et al.* 1991). The 'crop' plants of most gardening herbivores are simply widespread, opportunistic algae that are able to support persistent grazing and find competitive refuge within gardens. In some cases, there is evidence of 'fine tuning', for example the optimal cropping behaviour of certain limpets and perhaps the confinement of *Gelidium micropterum* to *Patella cochlear* territories (above), but gardening remains basically an indirect consequence of contest competition.

4. DISCUSSION

It is clear from the data reviewed above that a significant proportion of aquatic plant–animal interactions occur between species with relatively similar generation times, in contrast to the situation pertaining in the terrestrial analogues. However, it is not at all clear that such interactions can be usefully viewed as coevolutionary events.

Adopting a selfish-gene viewpoint, the strongest candidates for coevolution would be the different loci in the nuclear genome. Such loci are strongly constrained by the process of cell division to replicate at the same rate, although molecular mechanisms exist which permit some differences in rate of replication among loci (see Dover (1988) for examples). In these tightly constrained systems, there is clearly a short-term evolutionary advantage for mutually beneficial alleles at different loci to become tightly linked to form 'coadapted gene complexes'. One might expect that the outcome of such selection would be the production of a genome consisting of one tightly linked gene complex (but see Turner 1967). At one level, this has been achieved in the sense that the genomes of extant species comprise loci that cooperate with one another. The persistence of recombination among loci, however, is clear evidence that there is some disadvantage associated with too close an association. If loci become too tightly linked (over connected) there would be a loss of evolutionary flexibility in the system with the concomitant danger of an increased probability of extinction, clearly shown for example by obligate clonal organisms (Hughes, 1989).

To maintain a stable, coevolved partnership between interacting species whose life cycles are not as closely constrained as genes within a genome, much stronger 'genetic feedback' is required. That is, the connectedness of such systems must be high. This, however, is a precarious situation, because it increases the sensitivity of the system to environmental change (see the example of *Lottia alveus* and *Zostera marina* reviewed above).

Of course it is possible for mutualisms to stabilize and assume paramount importance at higher taxonomic levels (Boucher *et al.* 1984). Quintessential

examples include the consortia of microbially derived organelles of eukaryotic cells (Margulis 1981) and the endosymbiosis between dinoflagellates and corals (Spencer Davies 1991; Smith, this symposium). Such evolutionarily stable endosymbioses originally involved at least one unicellular partner, which could complete its life history within the confines of the association. Stability becomes more tenuous when one partner cannot live entirely in association with the other. The ascoglossan–chlorophyte association, falling short of mutualism, probably represents the evolutionary limit to an endosymbiotic relationship between a herbivore and its multicellular food plant.

For coevolution, in the sense of reciprocal, stepwise evolution to occur, the balance between the benefit of the interaction in the short term and the greater fragility of the system in the longer term, due to increased connectedness, must be biased strongly in favour of the benefit. With the exception of some of the endosymbiotic mutualisms described above, the strength of genetic feedback needed to maintain the mutualism may result in the costs outweighing the benefits, with the result that the majority of existing species interactions fall short of coevolved mutualisms.

We are deeply indebted to George Branch, Bill DeMott, John Gilbert, Mark Hay, Mark Huntley, Gustav-Adolf Paffenhoffer, Bernabe Santelices, Peter Steinberg, Bob Steneck, Hank Vanderploeg and Gray Williams, who were unstinting in their generosity and whose fine work forms much of the substance of this review.

REFERENCES

Allan, J. D. 1976 Life history patterns in zooplankton. *Am. Nat.* **110**, 165–180.

Boucher, D. J., James, S. & Kerster, K. 1984 The ecology of mutualism. *A. Rev. Ecol. Syst.* **13**, 315–347.

Branch, G. M. 1975 Intraspecific competition in *Patella cochlear* Born. *J. Anim. Ecol.* **44**, 263–282.

Branch, G. M. 1981 The biology of limpets: physical factors, energy flow, and ecological interactions. *Oceanogr. mar. Biol. Ann. Rev.* **19**, 235–380.

Branch, G. M., Harris, J. M., Parkins, C., Bustamente, R. H. & Eekhout, S. 1991 Algal "gardening" by marine grazers: a comparison of the ecological effects of territorial fish and limpets. In *Plant–animal interactions in the marine benthos* (ed. D. M. John, S. J. Hawkins & J. H. Price), ch. 18. Systematics Association.

Brawley, S. H. 1991 Mesograzers. In *Plant–animal interactions in the marine benthos* (ed. D. M. John, S. J. Hawkins & J. H. Price), ch. 11. Systematics Association.

Brawley, S. H. & Adey, W. H. 1977 Territorial behaviour of threespot damselfish (*Eupomacentrus planifrons*) increases algal biomass and productivity. *Env. Biol. Fishes* **2**, 45–51.

Butman, C. A. 1987 Larval settlement of soft sediment invertebrates: the spatial scales of pattern explained by active habitat selection and the emerging role of hydrodynamical processes. *Oceanogr. mar. Biol. Ann. Rev.* **25**, 113–165.

Carpenter, R. C. 1986 Partitioning herbivory and its effects on coral reef algal communities. *Ecol. Monogr.* **56**, 345–363.

Clark, K. B. & Busacca, M. 1978 Feeding specificity and chloroplast retention in four tropical Ascoglossa, with a discussion of the extent of chloroplast symbiosis and the evolution of the order. *J. mollusc. Stud.* **44**, 272–282.

D'Antonio, C. 1985 Epiphytes on the rocky intertidal red alga *Rhodomela larix* (Turner) C. Agardh: negative effects on the host and food for herbivores? *J. exp. mar. Biol. Ecol.,* **86**, 197–218.

DeMott, W. R., Zhang, Q.-X. & Carmichael, W. W. 1991 Effects of toxic cyanobacteria and purified toxins on the survival and feeding of a copepod and three species of *Daphnia. Limnol. Oceanogr.* (In the press.)

DeMott, W. R. & Moxter, F. 1991 Foraging on cyanobacteria by copepods: responses to chemical defenses, and resource abundance. *Ecology.* (In the press.)

Dixon, J., Schroeter, S. C. & Kastendiek, J. 1981 Effects of the encrusting bryozoan *Membranipora membranacea* on the loss of blades and fronds by the giant kelp *Macrocystis pyrifera* (Laminariales). *J. Phycol.* **17**, 341–345.

Dover, G. A. 1988 Evolving the improbable. *Trends Ecol. Evol.* **3**, 81–84.

Ehrlich, P. R. & Raven, P. H. 1964 Butterflies and plants: a study in coevolution. *Evolution* **18**, 586–608.

Estes, J. A. & Steinberg, P. D. 1988 Predation, herbivory, and kelp evolution. *Paleobiology* **14**, 19–36.

Fenchel, T. 1974 Intrinsic rate of natural increase: the relationship with body size. *Oecologia, Berl.* **14**, 317–326.

Fogg, G. E. & Thake, B. 1987 *Algal cultures and phytoplankton ecology.* (269 pages.) Madison, Wisconsin: University of Wisconsin Press.

Futuyma D. J. & Slatkin, M. 1983 Introduction. In *Coevolution* (ed. D. J. Futuyma & M. Slatkin, pp. 1–13. Sunderland, Massachusetts: Sinauer Associates.

Gilbert, J. J. 1990 Differential effects of *Anabaena affinis* on cladocerans and rotifers: mechanisms and implications. *Ecology* **71**, 1727–1740.

Grant, S. W. F., Knoll, A. H. & Germs, G. J. B. 1991 Probable calcified metaphytes in the late Proterozoic nama group, Namibia: origin, diagenesis and implications. *J. Paleont.* **65**, 1–180.

Hahn, T. & Denny, M. 1989 Tenacity-mediated selectiv predation by oystercatchers on intertidal limpets and its role in maintaining habitat partitioning by 'Collisella' scabra and *Lottia digitalis. Mar. Ecol. Prog. Ser.* **53**, 1–10.

Hatcher, B. G. 1988 Coral reef primary productivity: a beggar's banquet. *Trends Ecol. Evol.* **3**, 106–111.

Hay, M. E. 1991 Seaweed chemical defenses: their role in the evolution of feeding specialization and in mediating complex interactions. In *Ecological roles for marine secondary metabolites; explorations in chemical ecology series* (ed. V. J. Paul). Ithaca, New York: Comstock Publishing Associates.

Hay, M. E. & Fenical, W. 1988 Marine plant–herbivore interactions: the ecology of chemical defense. *A. Rev. Ecol. System.* **19**, 111–145.

Hay, M. E., Duffy, J. E. & Pfister, C. A. 1987 Chemical defenses against different marine herbivores: are amphipods insect equivalent? *Ecology* **68**, 1567–1580.

Hay, M. E., Duffy, J. E. & Fenical, W. 1988*a.* Seaweed chemical defenses: among-compound and among-herbivore variance. *Proc. 6th Intern. Coral Reef Symp.* **3**, 43–48.

Hay, M. E., Renaud, P. E. & Fenical, W. 1988*b* Large mobile versus small sedentary herbivores and their resistance to seaweed chemical defenses. *Oecologia, Berl.* **75**, 246–252.

Hay, M. E., Duffy, J. E. & Fenical, W. 1990 Host-plant specialization decreases predation on a marine amphipod: an herbivore in plant's clothing. *Ecology* **71**, 733–743.

Horn, M. H. 1989 Biology of marine herbivorous fishes. *Oceanogr. mar. biol. Ann. Rev.* **27**, 167–272.

Hughes, R. N. 1986 *A functional biology of marine gastropods.* (245 pages.) London & Sydney: Croom Helm.

Hughes, R. N. 1989 *A functional biology of clonal animals.* (331 pages.) London & New York: Chapman & Hall.

Huntley, M., Sykes, P., Rohan, S. & Martin, V. 1986 Chemically-mediated rejection of dinoflagellate prey by the copepods *Calanus pacificus* and *Paracalanus parvus*: mechanism, occurrence and significance. *Mar. Ecol. Prog. Ser.* **28**, 105–120.

Huntley, M. E., Ciminiello, P. & Lopez, M. D. G. 1987 Importance of food quality in determining development and survival of *Calanus pacificus* (Copepoda: Calanoida). *Mar. Biol.* **95**, 103–113.

Janzen, D. H. 1980 What is coevolution? *Evolution* **34**, 611–612.

Jensen, K. R. 1983 Factors affecting feeding selectivity in herbivorous Ascoglossa (Mollusca: Opisthobranchia). *J. exp. mar. Biol. Ecol.* **66**, 135–148.

Kaufman, L. 1977 The three-spot damselfish: effects on benthic biota of Caribbean coral reefs. *Proc. 3rd Intern. Coral Reef Symp.* **1**, 559–564.

Lassuy, D. R. 1980 Effects of "farming" behavior by *Eupomacentrus lividus* and *Hemiglyphidodon plagiometopon* on algal community structure. *Bull. Mar. Sci.* **30**, 304–312.

Lewis, S. M. 1986 The role of herbivorous fishes in the organization of a Caribbean reef community. *Ecol. Monogr.* **56**, 183–200.

Lindberg, D. R. 1990 Morphometrics and the systematics of marine plant limpets (Mollusca: Patellogastropoda). *Proc. Michigan Morphometrics Worksh.* (ed. F. J. Rohlf & F. L. Bookstein). Special Publ. No. 2. Ann Arbor, Michigan: University of Michigan Museum of Zoology.

Mann, K. H. 1973 Seaweeds: their productivity and strategy for growth. *Science, Wash.* **182**, 975–981.

Margulis, L. 1981 *Symbiosis in cell evolution.* (419 pages.) San Francisco: W. H. Freeman and company.

Menge, J. L. 1975 *Effects of herbivores on community structure of the New England rocky intertidal region: distribution, abundance and diversity of algae.* (164 *pages.*) Ph.D. thesis, University of Harvard.

Munoz, M. & Santelices, B. 1989 Determination of the distribution and abundance of the limpet *Scurria scurra* on the stipes of the kelp *Lessonia nigrescens* in Central Chile. *Mar. Ecol. Prog. Ser* **54**, 277–285.

Nelson, W. G. 1979 An analysis of structural pattern in an eelgrass (*Zostera marina* L.) amphipod community. *J. exp. mar. Biol. Ecol.* **39**, 231–264.

Norton, T. A. & Manley, N. L. 1990 The characteristics of algae in relation to their vulnerability to grazing snails. In *Behavioural mechanisms of food selection* (ed. R. N. Hughes), pp. 461–478. NATO ASI Series, G20, Springer-Verlag.

Oswald, R. C., Telford, N., Seed, R. & Happey-Wood, C. M. 1984 The effect of encrusting bryozoans on the photosynthetic activity of *Fucus serratus* L. *Estuar. Coastal Shelf Sci.* **19**, 697–702.

Paine, R. T. & Vadas, R. L. 1969 The effect of grazing by sea urchins, *Strongylocentrotus* spp. on benthic algal populations. *Limno. Oceanogr.* **14**, 710–179.

Paul, V. J. & Van Alstyne, K. L. 1988 Use of ingested algal diterpenoids by *Elysia halimedae* Macnae (Opisthobranchia: Ascoglossa) as antipredator defenses. *J. exp. mar. Biol. Ecol.* **119**, 15–29.

Pfister, C. A. & Hay, M. E. 1988 Associational plant refuges: convergent patterns in marine and terrestrial communities result from differing mechanisms. *Oecologia, Berl.* **77**, 118–129.

Porter, K. G. 1977 The plant-animal interface in freshwater ecosystems. *Am. Sci.* **65**, 159–170.

Price, P. W., Bouton, C. E., Gross, P., McPherson, B. A., Thompson, J. N. 1980 Interactions among three trophic levels: influence of plants on interactions between insect herbivores and natural enemies. *A. Rev. Ecol. System.* **11**, 41–65.

Sammarco, P. W. 1983 Effects of fish and damselfish territoriality on coral reef algae. I. Algal community structure. *Mar. Ecol. Prog. Ser.* **13**, 1–14.

Seed, R. & O'Connor, R. J. 1981 Community organization in marine algal epifaunas. *A. Rev. Ecol. System.* **12**, 49–74.

Sousa, W. P. 1979 Experimental investigations of disturbance and ecological succession in a rocky intertidal algal community. *Ecol. Monogr.* **49**, 227–254.

Spencer-Davis, P. S. 1991 Endosymbiosis in marine cnidarians. In *Plant–animal interactions in the marine benthos.* (ed. D. M. John, S. J. Hawkins & J. H. Price), ch. 23. Systematics Association.

Stallard, M. O. & Faulkner, D. J. 1974 Chemical constituents of the digestive gland of the sea hare *Aplysia californica* I. Importance of diet. *Comp. Biochem. Physiol.* **49**, 25–35.

Steele, J. H. 1974 *The structure of marine ecosystems.* Oxford: Blackwell Scientific Publications.

Steinberg, P. D. 1988 The effects of quantitative and qualitative variation in phenolic compounds on feeding in three species of marine invertebrate herbivores. *J. exp. mar. Biol. Ecol.* **120**, 221–237.

Steneck, R. S. 1982 A limpet coralline alga association: adaptations and defenses between a selective herbivore and its prey. *Ecology* **63**, 507–522.

Steneck, R. S. 1983 Escalating herbivory and resulting adaptive trends in calcareous algal crusts. *Paleobiology* **9**, 44–61.

Steneck, R. S. 1985 Adaptations of crustose coralline algae to herbivory: patterns in space and time. In *Paleoalgology* (ed. D. Toomy & M. Nitecki), pp. 352–366. Springer-Verlag.

Steneck, R. S. 1986 The ecology of coralline algal crusts: convergent patterns and adaptive strategies. *A. Rev. Ecol. System.* **17**, 273–303.

Steneck, R. S. 1991 Plant–herbivore coevolution: a reappraisal from the marine realm and its fossil record. In *Plant–animal interactions in the marine benthos* (ed. D. M. John, S. J. Hawkins & J. H. Price), ch. 21. Systematics Association.

Stoner, A. W. 1980 Abundance, reproductive seasonality and habitat preferences of amphipod crustaceans in seagrass meadows of Apalachee Bay, Florida. *Contrib. Mar. Sci.* **23**, 67–77.

Sykes, P. F. & Huntley, M. E. 1987 Acute physiological reactions of *Calanus pacificus* to selected dinoflagellates: direct observations. *Mar. Biol.* **94**, 19–24.

Turner, J. R. 1967 Why does the genome not congeal? *Evolution* **21**, 645–656.

Vadas, R. L. 1990 Comparative foraging behavior of tropical and boreal sea urchins. In *Behavioural mechanisms of food selection* (ed. R. N. Hughes), pp. 479–514. NATO ASI Series, G20, Springer-Verlag.

Vanderploeg, H. A., Paffenhofer, C.-A. & Liebig, J. R. 1990 Concentration-variable interactions between calanoid copepods and particles of different food quality: observations and hypotheses. In *Behavioural mechanisms of food selection* (ed. R. N. Hughes), pp. 595–613. NATO ASI Series, G20, Springer-Verlag.

Vine, P. J. 1974 Effects of algal grazing and aggressive behaviour of the fishes *Pomacentrus lividus* and *Acanthurus sohal* on coral-reef ecology. *Mar. Biol.* **24**, 131–136.

Wellington, G. M. 1982 Depth zonation of corals in the Gulf of Panama: control and facilitation by resident reef fishes. *Ecol. Monogr.* **52**, 223–241.

Williams, A. H. 1990 The threespot damselfish: a non-carnivorous keystone species. *Am. Nat.* **116**, 138–142.

Williams, G. A. 1990 *Littorina mariae*: a factor structuring low shore communities? *Hydrobiologia* **193**, 139–146.

Williams, G. A. & Seed, R. 1991 Interactions between macrofaunal epiphytes and their host algae. In *Plant–animal interactions in the marine benthos* (ed. D. M. John, S. J. Hawkins & J. H. Price), ch. 9. Systematics Association.

Wing, B. L. & Clendenning, K. A. 1971 Kelp surfaces and associated invertebrates. *Beih. Nova Hedwigia* **32**, 319–341.

Discussion

N. Knowlton (*Smithsonian Tropical Research Institute, Balboa, Republic of Panama*). One of the biggest differences between marine and terrestrial ecosystems seems to be the relative lack of specialist herbivores in the sea. What are Professor Hughes' explanations for this apparent pattern?

R. N. Hughes. We favour the idea (Hay & Fenical 1988) that insects are able, with consistent accuracy, to locate specific host plants, whereas drifting propagules of aquatic invertebrates are not. Certain insects, notably aphids (J. H. Lawton, personal communication), also are liable to uncontrolled drift, but aphids reproduce profusely by apomictic parthenogenesis, vastly increasing the chance of the clonal genome reaching suitable host plants (Hughes 1989). This interpretation, however, should be viewed cautiously. Marine propagules are often able to postpone settlement if suitable substrata are not located. They are, moreover, strongly attracted to substratum-specific stimuli.

S. B. Malcolm (*Department of Biology, Imperial College, Silwood Park, U.K.*). Are differences in scale between marine and freshwater ecosystems likely to influence plant–herbivore interactions? For example, are water-soluble, plant chemical defences against aquatic herbivores more likely in small, freshwater ponds with fairly simple physical characteristics than in large volumes of seawater? Could differences in light penetration between marine and freshwater systems mean that metabolic pathways for the synthesis of antiherbivore chemical defences are constrained in different ways? Thus should we expect different kinds of defences in various aquatic systems that also differ from those of terrestrial systems?

R. N. Hughes. Size of the water body apparently does not constrain the effectiveness of water-soluble, chemical defences. In freshwater ponds, cyanobacteria produce endotoxins that are retained within the cell, whereas in the open sea dinoflagellates leak exotoxins into the water. Oceanic water masses retain their identity for some considerable time and local concentrations of dinoflagellate toxin can be sufficient to impair somatic and population growth of herbivorous copepods (Huntley *et al.* 1987).

Chemical defences of benthic macroalgae most obviously differ from those of higher plants in the lack of alkaloids, halogenation and the derivation of polyphenolics from phloroglucinol. Lack of alkaloids probably can be explained by nitrogen limitation, seawater generally being depauperate in this nutrient (Hay & Fenical 1988). Plant-like bryozoans, on the other hand, produce alkaloids (Carle & Christophersen 1980), but they have a nitrogen-rich diet. Halogenation is enigmatic; it occurs in bryozoans (Carle & Christophersen 1980) and perhaps may be concerned with antifouling properties. The central role of phloroglucinol in macroalgae may be a phylogenetic legacy. Because macroalgae are largely of intertidal or shallow subtidal habit, it is perhaps unlikely that light penetration would constrain the kinds of metabolic pathway involved in the production of secondary metabolites. The above ideas, however, are speculative and more data are needed to answer these interesting questions.

Reference

Carle, J. S. & Christophersen, C. 1980 Marine alkaloids. 2. Bromo-alkaloids from the marine byozoan *Flustra foliacea*. Isolation and structure elucidation. *J. org. Chem.* **45**, 1586–1589.

P. J. Grubb (*Botany School, University of Cambridge, U.K.*). Among land plants the ones with most notable defences against animals include both species that grow particularly slowly, e.g. semi-desert cacti, and species which grow particularly fast, e.g. spiny or poisonous gap-demanders in forest such as brambles and deadly nightshade. Professor Hughes' account suggests that among seaweeds chemical defences are well developed in the slowest-growing species but poorly developed or absent in the fast-growing 'gap-demanders'. Could he please comment on this difference?

R. N. Hughes. Growth rate and repellency are not strongly correlated among seaweeds. Phlorotannins are present not only in modestly growing fucoids, but also in rapidly growing kelps such as *Macrocystis pyrifera* (North 1971). Species of *Laurencia*, comparable in growth, may be strongly defended or lack defences altogether. Here, defensiveness is correlated with the risk of herbivory (Hay & Fenical 1988).

Defences are conspicuously absent among green seaweeds. Apart from certain strongly defended, perennial occupants of tropical reefs, chlorophytes are characteristically flimsy, opportunistic plants. They are temporary gap-fillers, of poor competitive ability and attractive to herbivores (Littler & Littler 1980). They grow fast and easily regenerate from fragments that accidentally escape digestion and become voided in the faeces of herbivores (Santelices & Ugarte 1987). Undefended chlorophytes, then, are equivalent to ephemeral or annual herbs. Defended, perennial gap-fillers, equivalent for example to brambles in forests, include certain kelps such as *Agarum cribrosum* (Dayton 1975). The situation probably is not fundamentally different between terrestrial and marine vegetation.

References

Dayton, P. K. 1975 Experimental studies of algal canopy interactions in a sea otter-dominated kelp community at Amchitka Island, Alaska. *Fish. Bull.* **73**, 230–237.

Littler, M. M. & Littler, D. S. 1980 The evolution of thallus form and survival strategies in benthic marine macroalgae: field and laboratory tests of a function form model. *Am. Nat.* **116**, 25–44.

North, W. J. (ed.). 1971 The biology of giant kelp beds (*Macrocystis*) in California. *Beih. Nova Hedwigia* **32**, 600.

Santelices, B. & Ugarte, R. 1987 Algal life-history strategies and resistance to digestion. *Mar. Ecol. Progr. Ser.* **35**, 267–275.

The influence of grazing on the evolution, morphology and physiology of plants as modular organisms

ERKKI HAUKIOJA

Laboratory of Ecological Zoology, Department of Biology, and the Kevo Subarctic Research Station, University of Turku, SF-20500 Turku, Finland

SUMMARY

Plants are modular organisms, i.e. they consist of repetitive multicellular units. The integrity of the plant is arranged by active meristems that hormonally suppress activity of other meristems. This basic design makes it possible for plants to have semi-independent, or totally independent, parts even within one structural individual. Accordingly, plant parts like ramets or branches may be qualitatively different because of developmental, environmental or genetical influences. They may respond to herbivory separately from other parts in the same structural individual.

The modular structure allows easy recovery after damage by herbivores. Simultaneously it may constrain possible functions and lead to seemingly non-optimal responses. Effects of herbivory on the very basic modular design must be limited. Instead, herbivory may function as an evolutionary force modifying regulation of plant structure and function, like location of meristems, and rules determining outcomes of interactions among meristems. Indirectly, herbivory may prevent evolution of more unitary plant individuals.

1. INTRODUCTION

General theories about plant–herbivore interactions (Feeny 1975, 1976; Rhoades & Cates 1976) largely concentrated on active, exclusive, defence, and more recently, on resource availability, or on ability of plants to recovery (see, for example, Coley *et al.* (1985); Mattson *et al.* (1987). Most of the theories were developed by zoologists, who seem to have been impressed by difficulties which herbivores meet when attacking plants. This has been thought to be important for the plant. Assuming genetic basis for resistance, evolution towards more efficient defences seems inevitable; there has to be a balance settled by costs of defence and benefits of avoiding herbivory.

However, plants are modular organisms and modularity leads to some consequences that are different to those in traditional, unitary organisms (see, for example, Jackson *et al.* (1985); Jerling (1985)). These include a number of features which plant–herbivore theories have considered only superficially, or not at all. For example, the unit to which costs and benefits of defence are calculated, traditionally is an individual. But unlike in animals, functional responses may not be expressed by individuals plants, but by parts of the individual. The very basic modular structure of plants may exclude genetic variation for some responses that we would predict to evolve simply because we know they would be beneficial for unitary organisms. Induced responses of branches, instead of trees, to defoliation or pruning (Tuomi *et al.* 1988; Långström *et al.* 1990) serve as examples. Second, the modular structure of plants is accompanied by a, from a

zoologist's point of view peculiar, partly sink-regulated photosynthesis (Wardlaw 1990; Marshall 1990) which may make it relatively easy for plants to increase intake of carbon. This, and an ability of plants to create new resource-catching organs (Watson 1986), makes the assumption of a fixed resource budget much less uncertain in plants than in unitary animals. Simultaneously, plants may become relatively insensitive to small losses of foliar biomass, and may recover relatively easily after herbivory. Third, the modular structure of plants allows some surprising responses, like increased plant quality due to browsing (Danell & Huss-Danell 1985), and may also lead to methodological difficulties in experimentation (Haukioja *et al.* 1990).

The other side of the problem of modularity refers to the effect of herbivory on evolution of morphological and physiological mechanisms in plants. That must be an integral part of any general plant–herbivore theory because plants are under similar evolutionary forces as animals. The problem is how the basic modular structure of plants affects the validity of evolutionary predictions (see Tuomi & Vuorisalo 1989; Eriksson & Jerling 1990). In this paper I try to elucidate some basic principles of the modular design of plants, and whether and how herbivory functions as a causative agent in moulding that design.

2. PLANTS AS MODULAR ORGANISMS

Any attempt to characterize plants generally as modular organisms falls short in front of the multitude of specific solutions. For instance, there is huge

Phil. Trans. R. Soc. Lond. B (1991) **333**, 241–247
Printed in Great Britain

variation how specific plant features (like physiological integration among ramets, or transport of nutrients) are arranged in different plant species or even populations (see, for example, van Groenendael & de Kroon (1990)). Instead of describing the multitude of solutions, I present a verbal model (modular interaction model (Haukioja 1990)) to explicitly describe such features in plant modular structure and functioning which may modify plant responses to herbivory. My views of plant responses to herbivory, as well as the treatment of modularity in this paper, have been strongly influenced by behaviours of the mountain birch, *Betula pubescens* spp. *tortuosa*, the species that I have been studying for almost two decades. A birch tree is composed of modular units but two characteristic features of the mountain birch may influence generality of the conclusions: first mountain birches usually have a huge reservoir of dormant meristems, so that, in practice, meristem limitation (see Watson 1986; Geber 1990) is unlikely to occur. Second, mountain birch is a polycormic bush and its ramets and branches are functionally, and perhaps also structurally, compartmentalized to the extent that they respond individually, for example, to defoliation treatments.

Although these features make behavioural options for the birch different to, for example, palms, they still show the degree of decentralization in structural individuals and allow some generalizations based on such features.

(a) Modularity as a morphological and as a functional concept

The concept modularity conveys two totally different views: a morphological and developmental concept, and composition of plants as conglomerates of partly independent functional units.

Modular organisms are composed of repetitive multicellular units, none of which is vital for the structural individual (Vuorisalo & Tuomi 1985). Modularity is originally a developmental and morphological concept (Hallé *et al.* 1978; White 1979; Harper 1981). A module contains, for example, a leaf, a shoot and a meristem. Depending on the size of the plant, and the phyllotactic system, all or some of the modules are directly connected by vascular systems.

The morphological definition of modularity does not necessarily emphasize those plant features that are critical for herbivory. Neither does the morphological definition help to understand how herbivory affects plant traits. From the viewpoint of survival in the presence of herbivory, an important point in modularity is that although morphologically fixed, modularity allows immense flexibility in a functional sense.

Regulation of plant integration takes place by hormones. They are produced by active meristems, and by means of hormones active meristems modify transport of nutrients. Suppressed meristems are dormant. The meristem that is responsible for causing the strongest hormonal gradient is the local centre of metabolic activity. Watson & Casper (1984) called the resulting functional unit an IPU, an integrated physiological unit. Such a unit is partially autonomous especially in carbon economy.

When the plant develops, activity of local meristems also changes. After the developmental phase of a module, its boundaries need not coincide with limits of an IPU. A particular module may belong to one IPU at one time and to another at another time. One IPU may cover one or several modules.

(b) Integration of modules and IPUs: ecological and evolutionary views

If an individual plant simultaneously is an aggregation of structural modules, and of functional IPUs, what causes the obvious integration of the individual? I propose that the very basic intermodular rules make integration very simple, almost automatic. The simplest alternative is to assume that the IPU at the best location has to get control over nearby, less optimally located, IPUs. In above-ground parts of a plant this means that (i) the meristem which has the best access to light (ii) remains or becomes metabolically active, (iii) creates high concentrations of hormones (especially auxins, see Matthysse & Scott (1984)) which (iv) both suppress nearby meristems and guide resources (water and nutrients) to that meristem. The coupling of high meristematic activity at favourable sites with production of pertinent hormones and transport of resources, incidentally creates integration among above-ground parts of the plant. Below-ground parts obviously follow analogous principles.

The genetic code for behaviour of an IPU is in genes. However, responses of an IPU cannot be directly genetically programmed. Instead, there have to be codes for alternative position-dependent behaviours for IPUS. Behaviours of an IPU have to take into account the position of the IPU within the plant (and therefore possible dominance of other, especially nearby, IPUs), and the external environment. Formation of a potential, and performance of an actual, IPU may be constrained by both. Consequently, modular, or IPU, integrations have to be programmed so that under normal conditions IPUs produce phenotypes that are locally optimal, or close to optimal, under the conditions that the IPU encounters.

This also means that there have to be genetically determined rules for achieving, maintaining and giving up dominance among competing IPUs. These rules make an aggregation of IPUs, the individual, behave in a way resembling behaviour of a true individual. Still, within the limits of these rules, there may be true competition among IPUs. The genetically constrained, but physiologically genuine competition among IPUs may be the way in which the integration and optimal, or close to optimal, functioning of the structural individual is arranged.

Although the relations among IPUs are dynamic, there exist some regularities. The most familiar is apical dominance (Hillman 1984), the phenomenon that, for example, in freely growing trees, the topmost meristem is the most active. It forms a strong sink and has access to adequate resources. Apical dominance offers an example of genetically based but environmentally modified traits. Variance in the strength of apical dominance (in addition to factors like dormancy and activity of meristems, internode length and branching angle) makes it possible to model a high number of different plant types (Sutherland & Stillman 1990). The strength of apical dominance has a hereditary basis (Kozlowski 1971) and therefore is susceptible to artificial or natural selection. Simultaneously, apical dominance is under strong influence of environmental factors, like nutrient availability (Kozlowski 1964; McIntyre 1977; Slade & Hutchings 1987). Enhanced nutrient levels increase ability of other meristems to activate their own IPUs. Therefore, in clonal herbs for instance, spread of the clone (determined by the growth rate of apical, primary branches) is relatively insensitive to environmental factors (Oinonen 1967 *a*, *b*; Hutchings 1988) but growth of secondary and lower level branches is strongly modified by availability of resources.

The behaviour of a structural individual consists of behaviour of its IPUs; they are the functional units. Consequently, phenotypic selection (consequences of natural selection as a causative, instead of statistical, concept; see Endler 1986) has to concentrate on behavioural responses of functional units, IPUs. If there is genetic intra-individual variability among IPUs, produced by somatic mutations, true evolution may follow within a structural individual. Assuming no somatic mutations, phenotypic selection within or among IPUs of an individual does not lead to genetic response because of a lack of additive genetic variance.

Naturally, inter-individual selection can function as a causative evolutionary agent. It operates via the ability of individual IPUs to grow, to reproduce and to survive. The more optimal are local responses of IPUs, the higher is the fitness of the genotype. Growth and reproduction of the whole structural individual (genet in many cases) is the sum of modular values, and survival of the plant also depends on survival of the modules, although not in a straightforward fashion.

3. HOW DOES MODULARITY MODIFY CONSEQUENCES OF HERBIVORY?

Plant modularity, including functional aspects like organization of the plant to spatially and temporally dynamic IPUs, and the partially sink-regulated photosynthesis, affect some aspects of plant–herbivore interactions. Next I deal briefly with the following topics: tolerance to loss of foliage, ability of plants to recover after damage and degree of phenotypic plasticity. In addition, modularity modifies usability of some methodological practices like calculations of the value of lost tissue, and effects of simulated herbivory. In addition, it offers explanations for some seemingly non-adaptive plant responses.

(*a*) *Tolerance to loss of foliage and*
(*b*) *ability of plants to recover after damage*

Growing plants usually shed the oldest leaves. This in part depends on shelf shading; new leaves shade

older ones and when the performance of old leaves becomes uneconomical, their resources are transported to younger tissues, and the old ones die. For the plant the importance of losing small amounts of foliage has to be evaluated against this background. Owing to presence of dormant meristems and sink regulation, new leaves may be easily built, and old ones can survive longer. Under moderate or good resource conditions, plants can compensate, even overcompensate, losses (Paige & Whitham 1987; Maschinski & Whitham 1989).

Accordingly, a plant may easily recover after herbivory assuming that the two preconditions for successful recovery exist: there are extra meristems and resources are available. The surprising ease of recovery largely depends on the same set of factors as above. A rearrangement of dominance hierarchies among IPUs automatically happens after disturbance caused, for example, by herbivores: IPUs at the best locations become activated and redirect resources to themselves. Sink regulation further helps the plant to recover: activated meristems create new sinks, mature source-leaves slow down the rate of senescence and may increase or prolong their photosynthetic activity.

(*c*) *Degree of phenotypic plasticity*

Alternative position-dependent behavioural rules of IPUs indicate that phenotypic plasticity in plants must be large. This has long been familiar, for example, in the ease by which plants form ecotypes. Different parts of a tree may be different for herbivores, and the same tree may, completely or locally, vary temporally in quality for herbivores due to developmental or environmental factors. For instance, juvenile trees, or parts of them, often contain spines although mature tissues do not. Furthermore, foliage quality for a herbivore in previously non-defoliated mountain birch trees may be more similar to other tree species than to mountain birches defoliated a couple of years earlier (Haukioja *et al.* 1988).

Therefore, in long-lived plants, phenotypic plasticity may reduce evolution via natural selection; temporarily perhaps also replace it. A single genotype simply may be able to create the necessary array of phenotypes. This is especially relevant for defensive functions in plant species in which the generation time is much longer than that of the herbivore. For such plants it is important to be able to phenotypically change enough to minimize risks of emergence of specialized genotypes of the main pests. As a curiosity, this seems also to be the way in which herbivores may adapt to temporal variability in host quality. The quality of mountain birches may be so variable and perhaps unpredictable within and among *Epirrita* generations that the moth manages best by corresponding phenotypic plasticity. Accordingly, although *Epirrita* broods are significantly different in most analyses concerning, e.g. growth rates, no significant brood × host-plant type interactions have been found (Ayres *et al.* 1987; K. Ruohomäki & E. Haukioja, unpublished information).

(d) Calculations of the value of lost tissue

All leaves do not have the same value for the plant. Earlier plant-defence theories assumed plant tissues to be protected in relation to their vulnerability (Feeny 1975; Rhoades & Cates 1976), or their value to the plant (McKey 1979). Leaf value presumably was equated with its photosynthetic performance. However, removing a fixed amount of foliage, of equal age, has similar consequences for the plant only if its carbon economy is totally integrated. In that case all the resources go to the general pool. Consequences of foliage removal may be modified, for example, by the role which the leaf has in feeding specific tissues. Harper (1989) has discussed the value of a leaf in relation to functions by the recipient tissue. Haukioja *et al.* (1990) showed that damage to an individual long shoot leaf of the mountain birch specifically affected growth of the potential shoot from the axillary bud of that leaf in the next year.

(e) Effects of simulated herbivory

To show the value of the approach viewing plants as populations of modules, and IPUs, I take a few examples of possible methodological problems in experiments trying to elucidate how trees respond to damage.

Damage to individual leaves leads to synthesis of compounds which spread to other parts of the plant (Green & Ryan 1972) via phyllotactic connections (see Watson & Casper 1984). Therefore rapid induced responses may not spread first to the physically nearest leaves, but to leaves that are phyllotactically closest.

Without any explicit analysis about tree organization, it has sounded totally reasonable to assume that they are individual trees which either respond or do not respond to damage. However, in the case of mountain birch, they were not trees that reacted to defoliation treatments, but ramets and branches (Haukioja & Neuvonen 1985; Tuomi *et al.* 1988).

Another confounding factor was traced when studying effects of previous defoliation to later susceptibility of the foliage to herbivores. Such experiments, or observations, were thought to test existence of induced defences. But surprisingly, previous defoliation, or browsing, in some cases caused the plants later to be better as diet for later herbivores (Niemelä *et al.* 1984; Williams & Myers 1984; Danell & Huss-Danell 1985; Roland & Myers 1987; Craig *et al.* 1988). Haukioja *et al.* (1990) offered a possible interpretation for those observations by assuming herbivores to be able to damage or to take over the regulative system of such plants. Manipulations of host plants are possible because the hormonally mediated and decentralized control of plant functions is more susceptible to external factors than centralized control by the nerve system in animals.

(f) Explanations for some seemingly non-adaptive plant responses

An example of a seemingly non-adaptive response to herbivory is the manner in which, trees like apples and birches, behave after vole or hare damage to their bark. If the bark is damaged around the stem, the tree dies because the isolated trunk and branch system can no longer feed roots with photosynthetic products. However, if the tree experiences more extensive damage, and the whole stem is cut, the tree probably recovers. The reason is that, unlike in the former case when the canopy was left intact, new meristems are activated in the stump, and the tree can continue its functions without exhausting reserves in the root system. This indicates that in the tree there is no 'programme' taking care of the benefits of an 'individual tree'; otherwise different behaviours of trees in the above cases do not seem reasonable. The tree is not a tightly integrated organism but a by-product of its parts. Still, under normal conditions, the tree function as an organized individual.

The modular design may also explain some other ostensibly problematic observations relating to defensive responses by trees. For instance, if insect-induced deterioration in foliage quality is an exclusive defence against defoliators, it is very strange that it is delimited it to some ramets or branches only; important defoliators are so mobile that they necessarily attack the whole tree. Such induced responses may not be true defences (see Haukioja & Neuvonen 1985), but at any case such trees do not respond as totally coordinated units but as largely independent sub-systems.

Young tissues are often strongly avoided by vertebrate herbivores, and the reason is often chemical, as in the case of twig resins against hares (Bryant 1981). This may protect juvenile trees. However, the same difference in quality may be found between young and older parts of a ramet. In that case young parts are avoided, but still lost because the hare cuts the branch although it only eats older less resinous parts.

4. DISCUSSION

(a) How does herbivory modify modularity?

Plants contain many toxins that are detrimental for most herbivores. Demonstrations of the potency of such compounds have strongly moulded our ways of thinking how plants deter herbivory. If concentrations of these compounds have a genetic basis, herbivores can function as selective agents and guide evolution of relevant plant defences. At that level plants may defend themselves, and chemical or physical quality of the plant may be strongly modified by herbivores.

On the other hand, the deep structure of plants as modular, incompletely integrated organisms cannot change abruptly, perhaps not at all, simply because there may not be suitable additive genetic variance in relevant traits. It might be as impossible to assume that such traits can change due to herbivory as to assume rabbits can evolve wings because they would be useful for escaping from predators. Herbivory obviously has no or only a very limited role as a factor modifying structural and functional features of plants as modular organisms. Therefore, much of what plants do, and how they respond to herbivory, may have explanations that have nothing to do with herbivory.

However, herbivores may modify regulation of the deep structure. In practice this is hard to prove but there are many examples in line with this logic. For instance, unlike in trees and forbs, meristems of grasses are beneath the growing structure. Therefore, grasses are very tolerant to grazing by large herbivores (McNaughton 1985). Herbivores are known to function as an ecological force that may strongly modify plant form, apical dominance and sex expression (Whitham & Mopper 1985). Whether they can function as evolutionary forces, too, remains open. The compartmentalization of plants to semi-isolated units may be an adaptation preventing spread of diseases (Shigo 1984). It remains unclear whether herbivores can function as selective agents determining the degree of compartmentalization.

Could plants be very different and still cope with herbivory? Perhaps not. Although it is easy to find adaptive interpretations for why, for instance, plants are compartmentalized, a complete change in plant functions might be very difficult because of herbivory. For two reasons it is hard to imagine that plants could be guided by a central, localized decision-making organ: plants have hard-walled cells and it is obviously impossible for them to build complex organs at least in the same way as it happens in animals. In addition, a fixed wiring system to connect the hypothetical central processing organ with peripheral organs of a hypothetical, unitary plant, might be very sensitive to damage. Because a sedentary organism can always be caught by a mobile predator, a central processing unit should be protected extremely effectively which might be difficult. It is furthermore unclear how easily hypothetical organisms with such an organization could be able to recover after damage. Shifting dominance relations among modules, activation of previously dormant meristems and sink regulation are ways to make construction of new parts almost automatic when old parts are damaged and there are resources for recovery.

Summarizing, although the basic functions in plants cannot be easily changed by herbivory, herbivory can be an important factor preventing or making it difficult for plants to develop totally new designs: the new designs would have to be completed before they would be effective against herbivores. Still herbivores may both modify the regulative systems and therefore affect evolution of plant functions, and more directly select for features protecting the plant.

(b) Implications of plant modularity for theories about plant–herbivore interactions

The major task of general plant–herbivore theories is to explain existence of plants in spite of herbivores. The explanation why plants are able to do so does not only concern evolution of plant defences, but refers to several possible factors like: (i) herbivore numbers are low for reason unrelated to plants, (ii) most herbivores, especially insects do not recognize a specific plant as a potential host due to limitations of their host-recognition system, (iii) plants tolerate small-scale herbivory well because of features relating to

their modular structure, (iv) plants may successfully recover after strong herbivory, and (v) plants defend themselves either by toxins or by low availability of nutrients.

The classic plant–herbivore theories emphasized chemical plant defences and especially the need of plants to defend themselves differently against specialized and generalized herbivores (Feeny 1975; Rhoades & Cates 1976). More recent theories have stressed importance of resources (Coley *et al.* 1985), plant ecological strategies like growth forms (Mattson *et al.* 1987), and ability to recover (van der Mejden *et al.* 1988). All these theories rest on an assumption that plants behave optimally against herbivores. I have earlier proposed that although this assumption is generally valid, the modular organization of plants may in some cases modify plant responses to an extent reversing predictions based on *a priori* logic (Haukioja 1990).

At the deepest level we may have problems relating to the fact that evolutionary theorists have only recently started to appreciate the complications produced by phenotypic selection simultaneously functioning at different hierarchical levels, from modules to genets (Tuomi & Vuorisalo 1989; Eriksson & Jerling 1990). Consequently, theories of plant defences take modularity only partially into account as a possible modifier of how evolution by natural selection takes place (see, for example, Buss (1983, 1987); Whitham & Slobodchikoff (1981)).

A further evolutionary difficulty is exemplified by sink-regulated photosynthesis. If it is possible for plants to increase the photosynthetic rate of their leaves, why do they not do so always? Whatever is the correct explanation for this question, much theorizing rests on the importance of excess carbon in relation to nutrients without understanding connections between the physiological and evolutionary mechanisms. For example, Rhoades (1979) assumed photosynthesis to be sensitive to high levels of carbon-based defences, which in turn would prevent extreme concentrations of phenols in photosynthesizing tissues. However, at least in birch well 'defended' trees (rich in phenolic compounds, poor in sugars and nitrogen) had a higher photosynthetic rate than less well 'defended' birches (Prudhomme 1982). Furthermore, the role of nutritional factors, in addition to toxins, has played a minor role in general plant herbivore theories, opposite to, for example, German forest entomological literature (for examples, see Schwenke (1968); Schopf (1981)). The partially sink-regulated photosynthesis may have repercussions at this level, too. For instance, plants recovering from defoliation are often rich in phenols, are poor resources for herbivores, and may thereby defend themselves. However, because there is a strong negative correlation between phenol and sugar levels (Jensen 1988), it is also possible that poor performance of herbivores on such diets results from low sugar levels. And their levels may be low simply because sugars are immediately conveyed from leaves to the strong sinks created by meristems in recovering tissues.

As discussed above, herbivores may be able to modify plants relatively easily, making them more suitable as food, by disturbing intermodular interactions. Understanding such relations points to the importance of basing plant–herbivore theories explicitly on structural and functional mechanisms in plants. Simultaneously it calls for caution in adaptive interpretations: successful manipulation of plants' regulative systems can lead to consequences that are opposite to defences (see, for example, Niemelä *et al.* (1984); Danell & Huss-Danell (1985); Haukioja *et al.* (1990)). If that is possible, convincing demonstrations of true defences are hard to obtain.

REFERENCES

Ayres, M. P., Suomela, J. & MacLean, S. F. Jr. 1987 Growth performance of *Epirrita autumnata* (Lepidoptera: Geometridae) on mountain birch: trees, broods, and tree-brood interactions. *Oecologia, Berl.* **74**, 450–457.

Bryant, J. P. 1981 Phytochemical deterrence of snow-shoehare browsing by adventitious shoots of four Alaskan trees. *Science, Wash.* **213**, 889–890.

Buss, L. W. 1983 Evolution, development, and the units of selection. *Proc. natn. Acad. Sci. U.S.A.* **80**, 1387–1391.

Buss, L. W. 1987 *The evolution of individuality.* Princeton University Press.

Coley, P. D., Bryant, J. P. & Chapin, F. S. III 1985 Resource availability and plant antiherbivore defense. *Science, Wash.* **230**, 895–899.

Craig, T. P., Price, P. W. & Itami, J. K. 1986 Resource regulation by a stem-galling sawfly on the arroyo willow. *Ecology* **67**, 419–425.

Danell, K. & Huss-Danell, K. 1985 Feeding by insects and hares on birches earlier affected by moose browsing. *Oikos* **44**, 75–81.

Eriksson, O. & Jerling, L. 1990 Hierarchical selection and risk spreading in clonal plants. In *Clonal growth in plants: regulation and function* (ed. J. M. van Groenendael & H. de Kroon), pp. 79–94. The Hague: SPB Academic Publishing.

Endler, J. A. 1986 *Natural selection in the wild.* Princeton University Press.

Feeny, P. 1975 Biochemical coevolution between plants and their insect herbivores. In *Coevolution of animals and plants* (ed. L. E. Gilbert & P. H. Raven), pp. 3–19. Austin: University of Texas Press.

Feeny, P. 1976 Plant apparency and chemical defence. *Rec. Adv. Phytochem.* **10**, 1–40.

Geber, M. 1990 The cost of meristem limitation in *Polygonium arenastrum*: negative genetic correlations between fecundity and growth. *Evolution* **44**, 799–819.

Green, T. R. & Ryan, C. A. 1972 Wound-induced proteinase inhibitor in plant leaves: A possible defence mechanism against insects. *Science, Wash.* **175**, 776–777.

Groenendael, J. M. van & de Kroon, H. (eds) 1990 *Clonal growth in plants: regulation and function.* The Hague: SPB Academic Publishing.

Hallé, F., Oldeman, R. A. A. & Tomlinson, P. B. 1978 *Tropical trees and forests; an architectural analysis.* Berlin: Springer Verlag.

Harper, J. L. 1981 The concept of population in modular organisms. In *Theoretical ecology* (ed. R. M. May), pp. 53–77. Sunderland, Massachusetts: Sinauer Associates.

Harper, J. L. 1989 The value of a leaf. *Oecologia, Berl.* **80**, 53–58.

Haukioja, E. 1990 Toxic and nutritive substances as plant defence mechanisms against invertebrate herbivores. In *Pest, pathogens and plant communities* (ed. J. Burdon & S. Leather), pp. 219–231. Oxford: Blackwell Scientific Publications.

Haukioja, E. & Neuvonen, S. 1985 Induced long-term resistance of birch foliage against defoliators, defensive or incidental? *Ecology* **66**, 1303–1308.

Haukioja, E., Neuvonen, S., Hanhimäki, S. & Niemelä, P. 1988 The autumnal moth *Epirrita autumnata* in Fennoscandia. In *Dynamics of forest insect populations* (ed. A. A. Berryman), pp. 163–178. New York: Plenum.

Haukioja, E., Ruohomäki, K., Senn, J., Suomela, J. & Walls, M. 1990 Consequences of herbivory in the mountain birch (*Betula pubescens* ssp *tortuosa*): importance of the functional organization of the tree. *Oecologia, Berl.* **82**, 238–247.

Hillman, J. R. 1984 Apical dominance. In *Advanced plant physiology* (ed. M. B. Wilkins), pp. 127–148. London: Pitman.

Hutchings, M. J. 1988 Differential foraging for resources and structural plasticity in plants. *Trends Ecol. Evol.* **3**, 200–204.

Jackson, J. B. C., Buss, L. W. & Cook, R. E. (ed.) 1985 *Population biology and evolution of clonal organisms.* Yale University Press.

Jensen, T. S. 1988 Variability of Norway spruce (*Picea abies* L.) needles; performance of spruce sawflies (*Gilpinia hercyniae* Htg.). *Oecologia, Berl.* **77**, 313–320.

Jerling, L. 1985 Are plants and animals alike? A note on evolutionary plant population ecology. *Oikos* **45**, 150–153.

Kozlowski, T. T. 1964 Shoot growth in woody plants. *Bot. Rev.* **30**, 335–392.

Kozlowski, T. T. 1971 *Growth and development of trees*, vol. 1. New York: Academic Press.

Långström, B., Tenow, O., Ericsson, A., Hellqvist, C. & Larsson, S. 1990 Effects of shoot pruning on stem growth, needle biomass, and dynamics of carbohydrates and nitrogen in Scots pine as related to season and tree age. *Can. J. For. Res.* **20**, 514–523.

Marshall, C. 1990 Source-sink relations of interconnected ramets. In *Clonal growth in plants: regulation and function* (ed. J. M. van Groenendael & H. de Kroon), pp. 23–41. The Hague: SPB Academic Publishing.

Maschinski, J. & Whitham, T. G. 1989 The continuum of plant responses to herbivory: the influence of plant association, nutrient availability, and timing. *Am. Nat.* **134**, 1–19.

Matthysse, A. G. & Scott, T. K. 1984 Functions of hormones at the whole plant level of organization. In *Encyclopedia of plant physiology* (ed. T. K. Scott), new series **10**, 219–243. Berlin: Springer Verlag.

Mattson, W. J., Lawrence, R. K., Haack, R. A., Herms, D. A. & Charles, P.-J. 1987 Defensive strategies of woody plants against different insect feeding guilds in relation to plant ecological strategies and intimacy of association with insects. In *Mechanisms of woody plant defenses against insects; search for pattern* (ed. W. J. Mattson, J. Levieux & C. Bernard-Dagan), pp. 3–38. New York: Springer Verlag.

McIntyre, G. I. 1977 The role of nutrition in apical dominance. *Symp. Soc. Exp. Biol.* **31**, 251–273.

McKey, D. 1979 The distribution of secondary compounds within plants. In *Herbivores. Their interaction with secondary plant metabolites* (ed. G. A. Rosenthal & D. H. Janzen), pp. 55–133. New York: Academic Press.

McNaughton, S. J. 1985 Grazing lawns: animals in herds, plant form, and coevolution. *Am. Nat.* **124**, 863–886.

Meijden, E. van der, Wijn, M. & Verkaar, H. J. 1988 Defence and regrowth alternative plant strategies in the struggle against herbivores. *Oikos* **51**, 355–363.

Niemelä, P., Tuomi, J., Mannila, R. & Ojala, P. 1984 The

effect of previous damage on the quality of Scots pine foliage as food for Diprionid sawflies. *Z. angew. Entomol.* **98**, 33–43.

Oinonen, E. 1967*a* Sporal regeneration of bracken in Finland in the light of the dimensions and age of its clones. *Acta for. fenn.* **83** (1), 3–96.

Oinonen, E. 1967*b* The correlation between the size of Finnish bracken (*Pteridium aquilinum* (L.) Kuhn) clones and certain periods of site history. *Acta for. fenn.* **83** (2), 1–51.

Paige, K. N. & Whitham, T. G. 1987 Overcompensation in response to mammalian herbivory; the advantage of being eaten. *Am. Nat.* **129**, 407–416.

Prudhomme, T. 1982 The effect of defoliation history on photosynthetic rates in mountain birch. *Rep. Kevo Subarctic Res. Stat.* **18**, 5–9.

Rhoades, D. F. 1979 Evolution of plant chemical defense against herbivores. In *Herbivores. Their interaction with secondary plant metabolites* (ed. G. A. Rosenthal & D. H. Janzen), pp. 3–54. New York: Academic Press.

Rhoades, D. F. & Cates, R. G. 1976 Toward a general theory of plant antiherbivore chemistry. *Rec. Adv. Phytochem.* **10**, 168–213.

Roland, J. & Myers, J. H. 1987 Improved insect performance from host-plant defoliation: winter moth on oak and apple. *Ecol. Entomol.* **12**, 409–414.

Shigo, A. L. 1984 Compartmentalization: a conceptual framework for understanding how trees grow and defend themselves. *A. Rev. Phytopath.* **22**, 189–214.

Schopf, R. 1981 Quantitative Untersuchungen zur Nahrungsaufnahme und -ausnutzung der Blattwespe *Gilpinia hercyniae*. *Z. angew. Entomol.* **92**, 137–149.

Schwenke, W. 1968 Neue Hinweise auf eine Abhängigkeit der Vermehrung blatt- und nadelfressender Forstinsekten vom Zuckergehalt ihrer Nahrung. *Z. angew. Entomol.* **61**, 365–369.

Slade, A. J. & Hutchings, M. J. 1987 The effects of nutrient availability on foraging in the clonal herb *Glechoma hederacea*. *J. Ecol.* **75**, 95–112.

Sutherland, W. J. & Stillman, R. A. 1990 Clonal growth: insights from models. In *Clonal growth in plants: regulation and function* (ed. J. M. van Groenendael & H. de Kroon), pp. 95–111. The Hague: SPB Academic Publishing.

Tuomi, J., Niemelä, P., Rousi, M., Sirén, S. & Vuorisalo, T. 1988 Induced accumulation of foliage phenols in mountain birch: branch response to defoliation? *Am. Nat.* **132**, 602–608.

Tuomi, J. & Vuorisalo, T. 1989 Hierarchial selection in modular organisms. *Trends Ecol. Evol.* **4**, 209–213.

Vuorisalo, T. & Tuomi, J. 1986 Unitary and modular organisms: criteria for ecological division. *Oikos* **47**, 382–385.

Wardlaw, I. F. 1990 The control of carbon partitioning in plants. *New Phytol.* **116**, 341–381.

Watson, M. A. 1986 Integrated physiological units in plants. *Trends Ecol. Evol.* **1**, 119–123.

Watson, M. A. & Casper, B. B. 1984 Morphogenetic constraints on patterns of carbon distribution in plants. *A. Rev. Ecol. Syst.* **15**, 233–258.

White, J. 1979 The plant as a metapopulation. *A. Rev. Ecol. Syst.* **10**, 109–145.

Whitham, T. G. & Mopper, S. 1985 Chronic herbivory: Impacts on architecture and sex expression of Pinyon pine. *Science, Wash.* **228**, 1089–1091.

Whitham, T. G. & Slobodchikoff, C. N. 1981 Evolution by individuals, plant-herbivore interactions, and mosaics of genetic variability: the adaptive significance of somatic mutations in plants. *Oecologia, Berl.* **49**, 287–292.

Williams, K. S. & Myers, J. H. 1984 Previous herbivore attack of red alder may improve food quality for fall webworm larvae. *Oecologia, Berl.* **63**, 166–170.

Discussion

P. J. GRUBB (*Botany School, University of Cambridge, U.K.*). It seems to me that the ability of plants to produce regrowth shoots from dormant buds, which is so important in their response to grazing, would have evolved in a world without animals as a result of selection for vegetative survival of drought, fire, frost and wind damage. Similarly the production of many secondary chemicals in greater concentrations in regrowth shoots would probably have evolved in a world without animals as a result of selection for survival of attack by pathogenic microorganisms. In contrast the production of spines on regrowth shoots and not on 'normal' shoots of many species does suggest that specific selection by animals has been important. Does Dr Haukioja agree?

E. HAUKIOJA. I agree. However, animals obviously have also selected higher concentration of some secondary compounds, and presumably synthesis of some new compounds.

V. BROWN (*Imperial College, Silwood Park, Berkshire, U.K.*). Does Dr Haukioja have any idea what effect the various types and levels of foliar damage have on the root system?

E. HAUKIOJA. I do not have any real information.

W. J. BOND (*Botany Department, University of Cape Town, South Africa*). I would like to question the validity of the modular 'rules'. With the first rule 'meristems with the best ability to grow gain control over other meristems', one would predict trees shaped like solar collectors with no modules developing on the shady north-facing side of the canopy. But trees have symmetrical canopies. Is this not inconsistent with the modular argument?

E. HAUKIOJA. I assume that that would happen if 'gaining control of other meristems' would mean total inability of all subdued meristems to develop. That is not what normally happens: the growth of some meristems is totally inhibited, whereas only lessened in some others.

Optimization of gut structure and diet for higher vertebrate herbivores

R. McN. ALEXANDER

Department of Pure and Applied Biology, University of Leeds, Leeds LS2 9JT, U.K.

SUMMARY

A generalized herbivore gut is modelled as (i) a well-stirred anterior chamber in which microbial fermentation occurs; (ii) a tubular reactor in which digestion but no fermentation occurs; and (iii) a posterior fermentation chamber. The rate at which the herbivore gains metabolizable energy is calculated for diets that can be eaten at different rates and contain different energy densities of easily digested cell contents, and of cell wall materials that can be fermented but not digested. The optimum gut structure for each diet is determined. Chewing probably speeds digestion and fermentation but reduces eating time. Optimal chewing times are determined for particular diets and guts.

Herbivores often have a choice between poorer food that can be eaten fast and richer food that can only be eaten more slowly. Energy costs may be incurred in travelling between patches of the richer food. Optimal diet choices are predicted for herbivores with particular gut structures.

1. INTRODUCTION

This paper tackles two main questions. What is the optimum gut structure for a herbivore eating a specified diet? And what is the optimum diet for a herbivore with specified gut structure? Such questions have often been asked before: for recent examples see Penry & Jumars (1987), Verlinden & Wiley (1989), Hume (1989), Prins & Kreulen (1990) and Murray (1991). The novel feature of this paper is its use of a simple mathematical model that predicts the energy gain of a herbivore, given the composition of the diet, the rate at which it is eaten and the volumes of the principal segments of the gut.

Penry & Jumars (1986, 1987) showed how the theory of chemical reactors can help us to understand the design of digestive systems. Plug flow reactors (PFRs) are tubes in which reagents flow with negligible mixing along the length of the tube. Continuous flow stirred tank reactors (CSTRs) are tanks in which the reagents are kept thoroughly mixed. The concentrations of reagents fall gradually along the length of a PFR, but reagents entering a CSTR are diluted immediately to the concentrations at which they will leave it. For that reason, PFRs give higher yields in reactions in which any catalysts are added at entry. However, microbial populations cannot sustain themselves in PFRs, and would soon be washed out. Penry & Jumars (1986) suggested that the fermentation chambers in herbivore guts should be modelled as CSTRs, and the rest of the gut as a PFR. Herbivorous mammals, birds (Herd & Dawson, 1984) and reptiles (Troyer, 1984) have fermentation chambers in their guts.

Penry & Jumars (1987) used the Michaelis–Menten equation to predict rates of digestion by the herbivore's own enzymes, and the Monod equation to predict fermentation rates. I will make the simpler assumption that rates of digestion and fermentation are limited only by the quantities of substrates present, and proceed according to first-order kinetics. This assumption would be unsatisfactory if the animal were taking large meals at long intervals, so that the microbial population had to build up after each meal, but seems adequate for the steady-state model that will be developed. Waldo *et al.* (1972) and Mertens & Ely (1982) also assumed first-order kinetics in their models of the rumen.

2. THE MODEL

We will consider an idealized gut consisting of an initial CSTR (a rumen), representing a fraction v_1 of total gut volume: a PFR (fractional volume v_2): and a second CSTR (a caecum or colon, v_3) (figure 1).

The animal will be assumed to feed continuously, at a constant rate. This rate will be described by the dilution rate D, the volumetric rate of food intake divided by the total volume of the gut. We will assume (as Penry & Jumars (1987) also did) that the volume of the food remains unchanged as it passes through the gut. This implies that it spends on average times v_1/D, v_2/D and v_3/D in the three segments of the gut, and that the mean residence time for the entire gut is $1/D$. The concentrations of substances and microbes in the food and in the gut will be expressed as energy densities (heat of combustion per unit volume). Subscripts will be used to distinguish energy densities in the diet (subscript 0) from those in the three segments of the gut (1, 2 and 3, see figure 1).

Two groups of substrates will be distinguished. Sugars, starch, oils, protein and other easily digestible

Phil. Trans. R. Soc. Lond. B (1991) **333**, 249–255
Printed in Great Britain

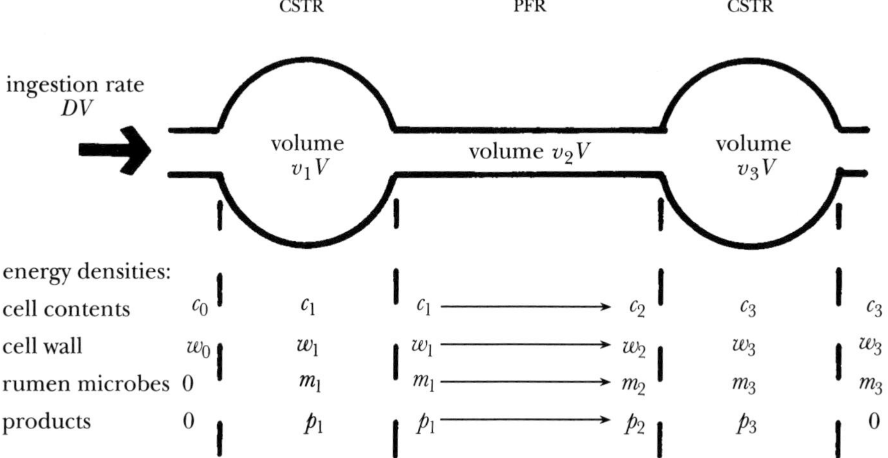

Figure 1. A diagram of the gut model.

materials will be referred to as 'cell contents' because that is their main origin. Their energy density is c_0 in the food, c_1 in the first CSTR, and so on. They can be fermented by microbes with fractional rate constant $r_{\mathrm{ferm,c}}$ or digested by the herbivore's own enzymes with rate constant r_{dig}. Cellulose and other materials that can be fermented but not digested are referred to as 'fermentable cell wall'. Their energy density in the food is w_0 and the rate constant for their fermentation is $r_{\mathrm{ferm,w}}$. This fraction excludes any of the cell wall polysaccharides that are protected from fermentation by their association with lignin.

The energy densities of rumen microbes (i.e. of microbes produced in the first CSTR) will be represented by the symbol m, with appropriate subscripts. That of absorbable products of digestion will be represented by p. We will assume that all these products are eventually absorbed but will not distinguish at any stage in the calculation between products that are still in the gut and those that have been absorbed into the blood-stream.

We will assume that fermentation of any substrate yields absorbable products and microbes whose heats of combustion are fractions Y_{ferm}, Y_{micr}, respectively of that of the substrate. Digestion of any substrate gives a fractional energy yield Y_{dig}.

We will now consider the segments of the gut in turn, starting with the first CSTR. Food travels through it (as through the rest of the gut) at a rate DV (dilution rate multiplied by gut volume). The energy density of cell contents is c_0 in the material entering the CSTR and c_1 in the material leaving, so the rate at which cell contents are broken down is $DV(c_0 - c_1)$. The volume of the CSTR is v_1V and cell contents are fermented at a fractional rate $r_{\mathrm{ferm,c}}$, so the rate of breakdown of cell contents is also $v_1Vc_1r_{\mathrm{ferm,c}}$.

$$DV(c_0 - c_1) = v_1Vc_1 r_{\mathrm{ferm,c}},$$
$$c_1 = Dc_0/(D + r_{\mathrm{ferm,c}} v_1). \tag{1}$$

Similarly, for fermentable cell wall materials

$$w_1 = Dw_0/(D + r_{\mathrm{ferm,w}} v_1). \tag{2}$$

The energy density of fatty acids and other absorbable products formed by these fermentations is

$$p_1 = Y_{\mathrm{ferm}}(c_0 - c_1 + w_0 - w_1), \tag{3}$$

and the energy density of microbes produced is

$$m_1 = Y_{\mathrm{micr}}(c_0 - c_1 + w_0 - w_1). \tag{4}$$

Now consider the PFR. The energy density of the remaining cell contents falls along its length from c_1 to c_2 as digestion proceeds at a rate r_{dig}. Each particle travels the length of the PFR in time v_2/D so

$$c_2 = c_1 \exp\left(-r_{\mathrm{dig}} v_2/D\right). \tag{5}$$

Cell wall materials (by our definition) cannot be digested, so

$$w_2 = w_1. \tag{6}$$

However, microbes that are carried out of the first CSTR with the food are digested here,

$$m_2 = m_1 \exp\left(-r_{\mathrm{dig}} v_2/D\right). \tag{7}$$

Thus, if no absorption had occurred through the gut wall, the energy density of absorbable products would increase to

$$p_2 = p_1 + Y_{\mathrm{dig}}(c_1 - c_2 + m_1 - m_2). \tag{8}$$

Finally, the food enters the second CSTR, where fermentation proceeds as in the first. Any microbes that arrive undigested will be dead and liable to fermentation at the same fractional rate as cell contents. Note that m_2, m_3 are energy densities of microbes originating in the first CSTR: microbes produced in the second will be lost in the faeces so there is no need for us to calculate their density. By the same arguments as for equations (1) to (3)

$$c_3 = Dc_2/(D + r_{\mathrm{ferm,c}} v_3), \tag{9}$$
$$w_3 = Dw_2/(D + r_{\mathrm{ferm,w}} v_3), \tag{10}$$
$$m_3 = Dm_2/(D + r_{\mathrm{ferm,c}} v_3), \tag{11}$$
$$p_3 = p_2 + Y_{\mathrm{ferm}}(c_2 - c_3 + w_2 - w_3 + m_2 - m_3). \tag{12}$$

The rate of flow through the gut is DV so the rate of formation of absorbable products in the entire gut is p_3DV. We will assume that these are completely absorbed (but see Dade *et al.* (1990)).

Equations (1) to (12) have been incorporated in a computer program that calculates $p_3 D$ for any specified diet (described by D, c_0 and w_0) and gut structure (v_1, v_2, v_3).

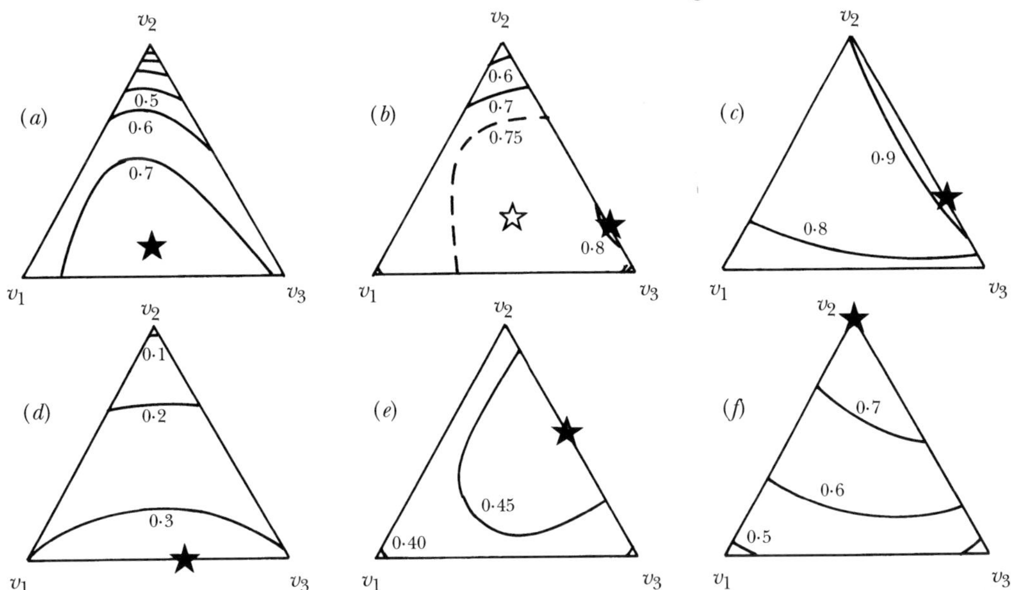

Figure 2. Rates of energy gain from specified diets in different guts. Each triangular graph represents the set of all possible gut structures (v_1, v_2, v_3), with contours representing rates of energy gain (expressed as $p_3/(c_0 + w_0)$). Filled stars represent global maxima and the hollow star a local maximum. Relative feeding rates D/r_{dig} are 0.05 in (a) to (c); and 0.5 in (d) to (f). The proportion of cell contents in the food, $c_0/(c_0 + w_0)$ is 0.1 in (a) and (d); 0.5 in (b) and (e); and 0.9 in (c) and (f).

3. VALUES FOR PARAMETERS

We will assume for lack of information that cell contents can be fermented or digested at equal rates: $r_{ferm,c} = r_{dig}$. However, we must assume that cell wall materials are fermented more slowly (Van Soest *et al.* 1988). We will generally assume $r_{ferm,w}/r_{dig} = 0.3$ but will also try other values.

Waldo *et al.* (1972) found that the mean fractional rate of fermentation of cellulose by rumen microbes from cattle was 0.07 h^{-1}. Mean retention times $(1/D)$ in the guts of herbivorous birds and mammals range from about 3 h to 130 h (Warner 1981). These data suggest that we should consider a wide range of values of D/r_{dig}: results will be presented for values ranging from 0.01 to 1.00. Values greater than 1.00 seem unlikely, as they would imply loss in the faeces of at least $1/e$ (37%) of cell contents.

The fractional yield of fermentation products Y_{ferm} will be taken to be 0.75 and the yield of microbes Y_{micr} will be taken to be 0.10. (Blaxter 1962). It will be assumed that no energy is lost in digestion: $Y_{dig} = 1.00$.

Figure 2 shows calculated energy gains for animals with different gut structures, on particular diets. Figure $2a, d$ show results for food containing a very low proportion of cell contents, lower than would be expected even in mature grass (Blaxter (1962) gives compositions by mass of grass and other foods from which compositions by energy content can be estimated). At a low rate of feeding (figure $2a$) the optimum gut has two large CSTRs ($v_1 \simeq v_2 \simeq 0.45$) and a very small intervening PFR ($v_2 \simeq 0.1$); and at a high rate of feeding (figure $2d$) it consists entirely of two CSTRs. (Two CSTRs are not equivalent to one large one because their contents do not intermix.)

Figure $2b, e$ shows results for food with a moderate proportion of cell contents, at the same two feeding rates. The optimum gut now has no first CSTR: at the lower feeding rate it has a small PFR and a large second CSTR (figure $2b$) but at the high rate it has a larger PFR (figure $2e$). Notice the local maximum in figure $2b$, close to the position of the global maximum in figure $2a$. The loss of the first CSTR from the optimum gut as food quality increases occurs abruptly at a bifurcation, when $c_0/(c_0 + w_0) = 0.45$ (at $D = 0.05$) or 0.47 (at $D = 0.5$).

Finally, figure $2c, f$ shows results for a diet with a very high proportion of cell contents, higher even than mangolds or grain (Blaxter 1962). At a low feeding rate, the optimum gut still consists of a small PFR and a large second CSTR (figure $2c$); but at a high rate it consists of a PFR and nothing else (figure $2f$).

There must be a maximum dilution rate for any CSTR, above which it cannot function as a fermentation chamber because microbial reproduction cannot keep pace with the dilution, and the microbial population is lost. The optimal gut for operation above this rate will consist of a PFR alone, for any food.

The results shown in figure 2 were all calculated for $r_{ferm,w}/r_{dig} = 0.3$. In other calculations, this ratio was given values of 0.2 and 0.5. The positions of the maxima were altered very little, with one exception: when figure $2b$ was recalculated for $r_{ferm,w}/r_{dig} = 0.5$, only one maximum was found, at the position of the local maximum in the figure. Increasing the ratio increases the relative feeding rate (D/r_{dig}) at which the bifurcation occurs.

In yet other calculations, the fractional yield of fermentation products Y_{ferm} was changed from the value used for figure 2 (0.75) to 0.50, with or without a further change of the microbial yield Y_{micr} from 0.10 to 0.25. These changes moved the optima only slightly, from the positions shown in figure 2.

Ruminant guts consist of a very large first CSTR (the reticulorumen), a small PFR (the abomasum and small intestine) and a second, smaller CSTR (the caecum and

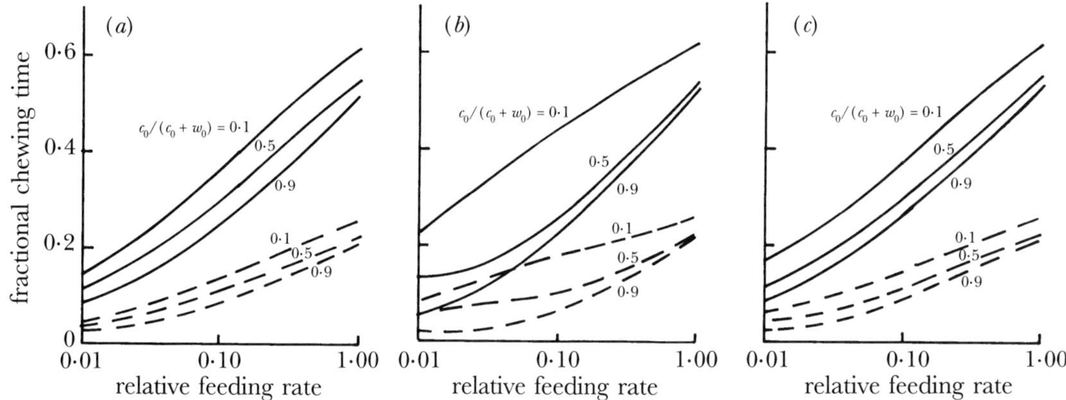

Figure 3. Graphs of optimal fractional chewing time against relative feeding rate D_{max}/r_{dig} for (a) a ruminant with $v_1 = 0.80$, $v_2 = 0.05$ and $v_3 = 0.15$; (b) a non-ruminant with $v_1 = 0$, $v_2 = 0.70$ and $v_3 = 0.30$; and (c) a non-ruminant with $v_1 = 0$, $v_2 = 0.30$ and $v_3 = 0.70$. The constant q is 1 (continuous lines) or 10 (broken lines).

colon): typical proportions, estimated from masses of gut contents, are $v_1 = 0.8$, $v_2 = 0.06$, $v_3 = 0.14$ (Maloiy *et al.* 1982). The model indicates that these proportions would give near-optimal energy gains from poor food (figure 2a, b). It suggests that a gut with a smaller rumen and a larger caecum and colon would be even better, but this suggestion may be misleading because the model takes no account of the reduction of volume that presumably occurs as the food travels along the gut, or the movement of food back and forth between rumen and mouth during rumination. Some ruminants eat poor food such as mature grasses but others select richer food (Jarman 1974).

Figure 2b, e show that a gut with no rumen but with a large posterior fermentation chamber (as in horses) may be optimal for diets containing moderate proportions of cell contents. This accords with Penry & Jumars' (1986) statement that horses outcompete ruminants when food is abundant and good. However, the model gives no support for Janis' (1976) contention that hindgut fermentation is better than a rumen for very poor as well as for rich diets.

Milton (1981) compared the monkeys *Alouatta*, which eats leaves, passing them through its gut relatively slowly; and *Ateles* which eats fruit, passing it through fast. The leaves presumably contain a moderate proportion of cell contents and the fruit a high one. Appropriately, as it seems from figure 2, the colon (the posterior fermentation chamber) is much smaller in *Ateles* than in *Alouatta*.

The model presented above takes no account of the habit of lagomorphs and some rodents, of eating faeces and so passing material through the gut for a second time.

5. OPTIMUM CHEWING TIME

In this section we ask, how long should a herbivore chew its food? Chewing breaks food up into small particles, damages cells and presumably speeds fermentation and digestion. There is some evidence from *in vitro* experiments that finely-ground food is fermented faster than larger pieces of the same food (Mertens & Ely 1982). However, the more an animal chews the less food, presumably, it has time to eat.

We do not know the functional relation between chewing time and rates of fermentation or digestion, but will assume that the rates r_{ferm} and r_{dig} are very low for unchewed food and are increased by chewing, approaching asymptotic values after long chewing times. Let τ be the fraction of the time available for eating or chewing, that is spent chewing, and let R (with appropriate subscripts) be the asymptotic value of a rate of fermentation or digestion. We assume

$$r(\tau) = R(1 - \exp(-q\tau/(1-\tau))), \tag{13}$$

where q is a constant: a high value of q would indicate that the asymptotic rate was approached rapidly. The effect of chewing time on energy gain was investigated by using $r(\tau)$ (with appropriate subscripts) instead of r in equations (1), (2), (5), (9) and (10). However r_{dig} in equation (7) and $r_{ferm,c}$ in equation (11) were left independent of chewing time, because they refer to the breakdown of microbes. We will assume $R_{ferm,c} = R_{dig} = r_{dig}$ and $R_{ferm,w} = 0.3\, r_{dig}$.

Account was taken of the reduced time available for eating by replacing D in equations (1) to (12) by

$$D(\tau) = (1-\tau)D_{max} \tag{14}$$

A computer program found the fractional chewing time that maximized the rate of energy gain $p_3 VD(\tau)$ for given gut structures and diets. Figure 3 shows that optimal chewing times are generally longer for foods that break down slowly when chewed ($q = 1$) than for foods that break down faster ($q = 10$). They are longer for foods that can be eaten rapidly (large values of D_{max}) than for those that can only be eaten slowly. They are also longer for foods that contain large proportions of fermentable cell wall materials. We should not be surprised that cattle chew for about 8 hours per day (Welch & Hooper 1988). The model implies that chewing occurs before swallowing, so tells us nothing about the special advantage of rumination.

Long chewing times imply rapid tooth wear, especially if the food contains abrasive particles, such as the silica particles in grass, or is contaminated by soil particles. Grazing mammals including cattle, horses and lagomorphs have hypsodont (high crowned) teeth.

Herbivorous birds grind food between stones in a muscular gizzard (see Herd & Dawson 1984) but the

Phil. Trans. R. Soc. Lond. B (1991)

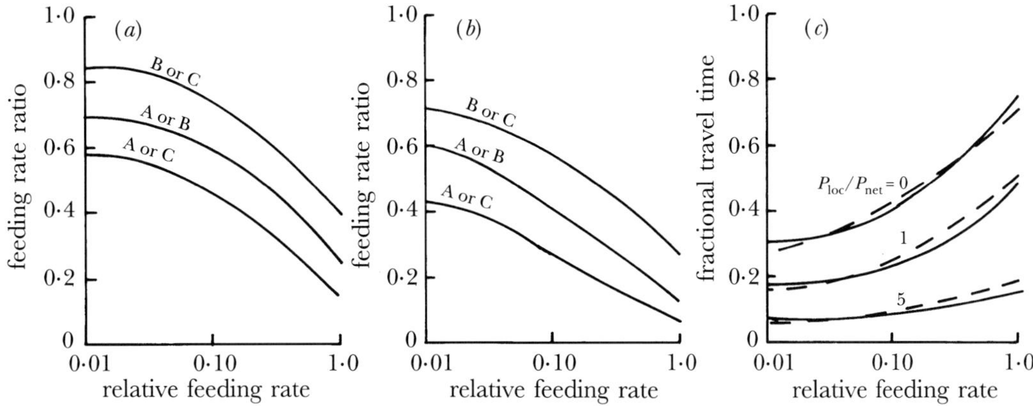

Figure 4. Graphs showing when a rich food that can be eaten only slowly gives the same rate of energy gain as a poor food that can be eaten fast, for the same two herbivores as in figure 3 *a, b*. (*a*) Breakeven feeding-rate ratios $D_{\text{rich}}/D_{\text{poor}}$ for the ruminant, for three pairs of foods, when energy costs of travel are negligible; (*b*) shows the same for the non-ruminant and (*c*) shows breakeven fractional travel time τ' for the ruminant choosing between foods A and B (continuous lines) and the non-ruminant choosing between B and C (broken lines). The energy densities of cell contents and fermentable cell wall materials in the foods are A, $c_0/E = 0.05$, $w_0/E = 0.50$; B, $c_0/E = 0.30$, $w_0/E = 0.50$; C, $c_0/E = 0.75$, $w_0/E = 0.20$.

arguments of this section do not apply to them because the storage capacity of the crop enables them to eat while grinding.

The quality of food (represented by $c_0/(c_0 + w_0)$) has much more effect on optimal chewing time for a herbivore with a small fermentation chamber (figure 3 *b*) than for those with larger ones (figure 3 *a, c*).

6. OPTIMIZATION OF DIET

Animals will generally gain energy faster by eating a richer diet (one containing a higher proportion of cell contents) if the alternative foods can be eaten at the same rate. This section examines choices between poor foods that can be eaten fast and richer food that can only be eaten more slowly, for example between eating grass indiscriminately in large mouthfuls or selecting the small but richer young shoots. Initially we will ignore the energy costs of any locomotion that may be involved in seeking out the richer food.

We will consider pairs of foods. In each case, rates of energy gain have been calculated for the poorer food, by using equations (1) to (12), for specified feeding rates D_{poor}. The feeding rate D_{rich} for the richer food, that gave the same rate of energy gain, was then found. Figure 4 shows breakeven feeding rate ratios $D_{\text{rich}}/D_{\text{poor}}$ for pairs chosen from three foods. Food (A) is intended to represent mature leaves (either grass or dicotyledons); (B) young leaves; and (C) very rich food such as fleshy fruit or grain (Blaxter 1962).

Figure 4 *a, b* shows that when feeding rates are low, the better quality of a rich food will compensate for only a small reduction in feeding rate. At higher feeding rates, cell wall materials are less completely fermented and breakeven feeding rate ratios are lower. This helps to explain two observations. First, grazing antelopes are less selective when feeding on low biomass pastures (Murray 1991). Secondly, very large bovids (which have relatively low mass-specific metabolic rates, and correspondingly low values of D) tend to be less selective than smaller ones (Jarman 1974). An

additional reason for large herbivores being less selective is that they can take larger bites, so their feeding rates may be reduced more, when they eat food that is available only in small mouthfuls.

At any particular value of D_{poor}, breakeven feeding-rate ratios are lower for the non-ruminant of figure 4 *b* than for the ruminant of figure 4 *a*; this particular non-ruminant ferments cell wall materials less completely than the ruminant, so gains more advantage from better food. The differences would be reduced if the non-ruminant were given a larger posterior CSTR. They show however that in some circumstances one herbivore should prefer the poorer but more plentiful of two foods, whereas another, even of the same size, should prefer the richer.

Thorns reduce the rate at which herbivores can eat a plant's leaves (Cooper & Owen-Smith 1986), so reduce the probability of their being preferred over other food.

The relative merits of diets may depend on the energy costs of feeding on them; for example, herbivores may have to travel further between bites of a richer but less common food. We will assume (unlike Murray 1991) that the rate of intake of food is reduced in proportion to the fraction τ' of the available time that is spent travelling

$$D = (1 - \tau')D_{\text{max}}. \tag{15}$$

This implies that the foods being compared could be eaten at the same rate D_{max}, if there were no need to travel. This is realistic for some pairs of foods but not (as discussed above) for others.

We will assume that metabolic power P_{loc} is required for locomotion, so that the net rate of energy gain by feeding (with its associated locomotion) is given by

$$P_{\text{net}} = p_3 DV - P_{\text{loc}}\tau'$$
$$= (1 - \tau')p_3 D_{\text{max}}V - P_{\text{loc}}\tau'. \tag{16}$$

We will again consider a pair of foods: a poorer, continuously distributed food that can be eaten with negligible travel costs and a richer dispersed food. A computer program has been used to find the fractional travel time for the richer food that makes P_{net} the same as for the poorer one.

Figure 4c shows results for the ruminant of figure 4a, choosing between foods A and B and for the non-ruminant of figure 4b choosing between foods B and C. The results for these two cases are almost identical. P_{loc} is expressed as a multiple of P_{net}; that is, approximately, as a multiple of metabolic rate. The graphs show that when $P_{loc}/P_{net} = 1$ or 5, the proportion of time, that the animal can spend travelling and still benefit from the richer food, is greatly reduced. $P_{loc}/P_{net} = 1$ represents approximately the energy cost of walking at speeds around 1 ms^{-1} (Taylor *et al.* 1982). P_{loc}/P_{net} might perhaps reach 5 if the animal ran between patches of food or if it had to climb trees to reach them.

REFERENCES

Blaxter, K. L. 1962 *The energy metabolism of ruminants.* London: Hutchinson.

Cooper, S. M. & Owen-Smith, N. 1986 Effects of plant spinescence on large mammalian herbivores. *Oecologia* **68**, 446–455.

Dade, W. B., Jumars, P. A. & Penry, D. L. 1990 Supply-side optimization: maximizing absorbtive rates. In *Behavioural mechanisms of food selection* (ed. R. N. Hughes), pp. 531–556. London: Springer Verlag.

Herd, R. M. & Dawson, T. J. 1984 Fiber digestion in the emu, *Dromaius novaehollandiae*, a large bird with a simple gut and high rates of passage. *Physiol. Zool.* **57**, 70–84.

Hume, I. D. 1989 Optimal digestive strategies in mammalian herbivores. *Physiol. Zool.* **62**, 1145–1163.

Janis, C. 1976 The evolutionary strategy of the Equidae and the origins of rumen and cecal digestion. *Evolution* **30**, 757–774.

Jarman, P. J. 1974 The social organisation of antelope in relation to their ecology. *Behaviour* **48**, 215–267.

Maloiy, G. M. O., Clemens, E. T. & Kamau, J. M. Z. 1982 Aspects of digestion and *in vitro* rumen fermentation rate in six species of East African wild ruminants. *J. Zool., Lond.* **197**, 345–353.

Mertens, D. R. & Ely, L. O. 1982 Relationship of rate and extent of digestion to forage utilization – a dynamic model evaluation. *J. Anim. Sci.* **54**, 895–905.

Milton, K. 1981 Food choice and digestive strategies of two sympatric primate species. *Am. Nat.* **117**, 496–505.

Murray, M. G. 1991 Accounting for diet quality in grazing ruminants. *J. Anim. Ecol.* (In the press.)

Penry, D. L. & Jumars, P. A. 1986 Chemical reactor analysis and optimal digestion. *Bioscience* **36**, 310–315.

Penry, D. L. & Jumars, P. A. 1987 Modelling animal guts as chemical reactors. *Am. Nat.* **129**, 69–96.

Prins, R. A. & Kreulen, D. A. 1991 Comparative aspects of plant cell wall digestion in mammals. In *The rumen ecosystem* (ed. S. Hoshino, R. Onodera, H. Minoto & H. Itabashi), pp. 109–120. Tokyo: Japan Scientific Societies Press.

Taylor, C. R., Heglund, N. C. & Maloiy, G. M. O. 1982 Energetics and mechanics of terrestrial locomotion. I. Metabolic energy consumption as a function of speed and body size in birds and mammals. *J. exp. Biol.* **97**, 1–21.

Troyer, K. 1984 Structure and function of the digestive tract of a herbivorous lizard *Iguana iguana. Physiol. Zool.* **57**, 1–8.

Van Soest, P. J., Sniffen, C. J. & Allen, M. A. 1988 Rumen dynamics. In *Aspects of digestive physiology in ruminants* (ed. A. Dobson & M. J. Dobson), pp. 21–42. Ithaca: Comstock.

Verlinden, C. & Wiley, R. H. 1989 The constraints of digestive rate: an alternative model of diet selection. *Evol. Ecol.* **3**, 264–273.

Waldo, D. R., Smith, L. W. & Cox, E. L. 1972 Model of cellulose disappearance from the rumen. *J. Dairy Sci.* **55**, 125–129.

Warner, A. C. I. 1981 Rate of passage of digesta through the gut of mammals and birds. *Nutr. Abs. Rev.* B **51**, 789–820.

Welch, J. G. & Hooper, A. P. 1988 Ingestion of feed and water. In *The ruminant animal: digestive physiology and nutrition* (ed. D. C. Church), pp. 108–116. Englewood Cliffs: Prentice-Hall.

Discussion

L. DE BRUYN (*Laboratory of General Zoology, University of Antwerp, Belgium*). Some groups of insects (e.g. the Diptera family Chloropidae) infest their host plant with phytopathogenic symbionts. The larvae damage the host-plant's tissues and inoculate them with the bacteria. This leads to the development of local bacteriosis in the injured tissues, with maceration and lysis effects, and a number of reactive changes in the plant's structures. The insects' larvae develop in a pool of disintegrating plant tissues saturated with nutritive bacteria. In the process of feeding, the larvae ingest the bacteria along with the plant-tissue juice and use them for food. In this way the larvae have developed a sort of external CSTR.

When extrapolating the model for plant food processing by higher vertebrates to invertebrates, this phenomenon has to be taken into consideration to resolve observed deviations from the expected digestive system.

R. McN. ALEXANDER. I have nothing to add to this interesting point.

R. N. HUGHES (*School of Biological Sciences, University College of North Wales, Bangor, U.K.*). By scaling for size, does Professor Alexander's model ignore constraints imposed by size itself? For example, a unit of cell wall material may take a certain time to be broken down by fermentation. The smaller the body, the shorter will be the residence time within an CSTR. Insects may be too small for an CSTR to be worthwhile. On the other hand, *Diplodocus*, with its huge CSTR could probably prolong fermentation long enough to make a nutritious broth even out of cycads!

R. McN. ALEXANDER. The model finds the optimum gut structure for a particular diet, but even the optimum gut may not break food down fast enough to support the high mass-specific metabolic rate of a very small herbivore. The model shows that guts consisting largely of CSTRs are optimal for poor diets containing a lot of cell wall and very little cell contents. Such a diet may be incapable of supporting a small herbivore with a gut consisting mainly of CSTRs, but with any other gut structure the herbivore would do even worse. Small herbivores may need fairly rich diets.

P. W. SKELTON (*Department of Earth Sciences, The Open University, Milton Keynes, U.K.*). Why does Professor Alexander have two CSTRs in his initial herbivorous gut model? Is this a theoretically predicted ideal arrangement?

R. McN. ALEXANDER. My model, consisting of two CSTRs and an intervening PFR (any of which may be omitted), is the simplest capable of imitating reasonably well the whole observed range of herbivore guts. Ruminants, for example, have a rumen (CSTR), a small intestine (PFR) and a hind gut (CSTR). A CSTR can always be improved by replacing it with two CSTRs in series (with the same total volume), so that substrate concentrations do not fall immediately to their final values. Some herbivore fermentation chambers (for example, the stomachs of kangaroos) are relatively long, so there may be little mixing between the contents of their proximal and distal ends, and they may perhaps function as several CSTRs in series, but this has not been shown.

I did not consider the possibility of two PFRs with an intervening CSTR, because I do not know of any animal with such a gut.

C. G. JONES (*Institute of Ecosystem Studies, Millbrook, New York, U.S.A.*). Under what circumstances and with what consequences would coprophagy (e.g. rabbits) or internal recycling (e.g. koalas) be advantageous, based on Professor Alexander's model?

R. McN. ALEXANDER. The model is concerned only with energy, so cannot take account of the possible value of re-ingested faeces or caecotrophs as sources of nitrogen and vitamins. Faeces will presumably have lower metabolizable energy contents than the food from which they are derived, so animals are unlikely to be able to gain energy by eating faeces in preference to other food. However, if part of the day is spent resting or hiding where fresh food is not available, any energy obtained by re-ingesting faeces then is pure gain. Caecotrophs could have higher metabolizable energy content than the food from which they are derived, if they are formed selectively from the more digestible constituents of the food.

C. G. JONES. Given that detoxification of plant secondary metabolites can be done by the herbivore and its symbiotic microorganisms, would it be possible to develop similar models for detoxification? Could such models be coupled with Professor Alexander's model?

R. McN. ALEXANDER. My model can take account of plant secondary metabolites, only in so far as they affect rates of digestion or fermentation of the food.

Evolution of insect morphology in relation to plants

ELIZABETH A. BERNAYS

Department of Entomology, and Department of Ecology and Evolutionary Biology, University of Arizona, Tucson, Arizona 85721, U.S.A.

SUMMARY

This short review points out some of the major physical problems faced by insects feeding on plants, and some of the kinds of morphological adaptations that have been noted to date. Major emphasis is given to two factors: the nature of the plant surface, and the difficulty of dealing with hard or tough food. The surface provides a great variety of terrains that require specialization for maximizing tenacity and agility, especially for small insects. It is suggested that natural enemies may provide significant selection for the relevant morphologies. The difficulty of feeding upon certain plant tissues is shown to be overcome in different ways by different herbivore groups. In the case of tough leaves for example, grasshopper mandible adaptations appear to have evolved to maximize efficiency of processing. On the other hand, in the case of caterpillars, mandible adaptations for tough food appear to be minimizing handling time with a concomitant reduction in risk of predation. Convergent evolution is shown in both grasshoppers and caterpillars for dealing with leaves of similar design and an example is given of rapid evolution of mouthpart morphology in response to differences in host characteristics. These examples are used to indicate the risks of using such characters in establishing phylogenetic relationships. Finally, it is pointed out that plant chemical qualities and plant ecological factors can influence insect morphological features.

1. INTRODUCTION

Compared with Darwin's finches, selection pressure on insect herbivore morphology in relation to food has been almost ignored. There are comparative morphological studies on such structures as tongue length in bees but evolutionary approaches are relatively recent. Yet phytophagous insects provide a wealth of variety that reveals physical problems and the consequent selective pressures that plants provide: the evolution of morphological adaptations has certainly been as important as the much discussed chemical ones. As structures can now be studied quantitatively we have a new opportunity for the study of morphological character evolution.

Plant-feeding insects have been the subject of many theoretical and experimental studies and the examples given here will provide a first brief attempt to consider morphology. Plants present a series of potential problems that could influence the evolution of insect form, including superficial impediments to progress or attachment, impediments to reaching the food substrate, hardness, toughness or abrasiveness, nutritional deficiency or toxicological stress, and temporal or spatial challenges in plant distribution.

It is also unlikely that the animal adaptations to these problems would be the same without the additional influence of organisms in other trophic levels. For example, the morphological adaptations involved in host-specific crypsis show the relation between herbivore morphology, their hosts and their predators. The third trophic level is of particular importance in the case of host-related adaptations of small herbivores such as insects.

2. SURFACES

Surface features are significant for all insects, and tarsal claws are of general value for grasping at irregularities and maintaining a hold on the substrate. However, it might be expected that the variety of adaptations in tarsi will be greatest in small insects for whom the terrain is relatively more varied and conspicuous. Plant surface features have indeed posed considerable challenges for the herbivores that live, rest and feed upon them. Southwood (1973) drew attention to this fact in his description of the 'attachment hurdle' that herbivores had to overcome, and more recently several different elegant adaptations for attachment to particular types of leaf surface have been described. For example, Stork (1980) showed that in chrysomelid beetles, as in geckos, molecular adhesion from the thousands of minute setae on each tarsal pulvillus are responsible for an ability to walk on glabrous leaves. Individuals could hold on against a force of nearly 50 times their own body mass, so that whiplash of leaves in a gusty wind does not dislodge them. By contrast, Lee *et al.* (1986) showed that certain leafhoppers in the genus *Empoasca* can produce a miniature suction cup with each tarsal pulvillus which enables them to effectively hold onto smooth surfaces. On contact with the surface the pulvillus assumes a circular shape and fluid fills the area circumscribed. The fluid secretion

Phil. Trans. R. Soc. Lond. B (1991) **333**, 257–264
Printed in Great Britain

and withdrawal is apparently under control and the system works very effectively on the preferred smooth host leaves, but is ineffective on hairy surfaces.

There are also adaptations for movement on hairy leaf surfaces. Southwood (1986) described special 'pseudoaroliae' on the tarsi of certain mirid bugs. Individuals are able to grasp hairs between this projection and the pretarsal claw, allowing a very effective grasp of the hair. A different development has been described for a specialist oak aphid, *Myzocallis schreiberi* (Kennedy 1986). Like other related aphids its tarsi have a pair of claws terminally and a pair of flexible spatulate empodia. These empodia are laid along and around surfaces and provide additional contact area, but in *M. schreiberi* the empodia are more flexible, allowing a good grasp on the hairs of the densely pubescent leaves of *Quercus ilex*. As a result it is much more effective at holding on than relatives that feed mainly on other, more glabrous oaks. Such mechanisms may be common among the many insect species that appear to prefer leaves with trichomes (Southwood 1986).

Moran (1986*b*) found a highly significant negative correlation between host trichome density and length of hind tarsi in *Uroleucon* species, with unrelated species groups from different geographical regions showing a similar relation. Moran suggests that short tarsi help prevent entanglement in trichomes whereas long ones provide a better grip on glabrous surfaces. By contrast, Eastop (1985) noted that aphids on hosts with sticky exudates had short tarsi; species described from the extremely resinous *Grindelia squarrosa* have no tarsi at all (Gillette 1911). Presumably 'walking on tip toe' reduces work and allows greater agility.

The ability to negotiate the surfaces of host plants with agility has value not just for feeding and ovipositing, but also for escaping from predators. For many herbivores the leaf surface is not just a feeding site and habitation, but the lifetime habitat where predators may be more abundant than prey. For example, the ability to jump away from a predator and successfully land without falling was an important component of survivorship of alfalfa aphids in the presence of their coccinellid predators (Bernays 1989). Moran (1986*a*) noted that in the aphid genus *Uroleucon*, the specialist species on goldenrod was also much better able to escape predation than was the generalist species. Indeed, it appears very likely that one of the benefits of a narrow host range is the possibility of developing specialized tarsal morphology that promotes agility on the host (Kennedy 1986), and thus efficient escape behaviours (Bernays 1989). In addition, herbivore fall-off rates can be 10% of the population per day, with the majority of those falling becoming prey for ground predators (Winder 1990). There must be intense selection pressure for competence on the host surface.

The nature of the surface is also important directly in feeding; the morphological challenges again relate to trichomes on the one hand and hard or ultrasmooth surfaces on the other. Hoffman & McEvoy (1985) found that the dense layer of simple lanate trichomes on the stems of certain host plants provided a problem

to the spittle bug, *Philaenus spumarius*. When the length of the trichomes exceeded the length of the rostrum, the tip of the rostrum could not contact the leaf surface. Long rostra are an advantage on such plants, indeed the adult *P. spumarius* has a relatively long rostrum (1 mm), and it is the younger nymphs with shorter rostra whose feeding sites are limited. An analogous study on aphids specialized for feeding on *Tilia* spp. with dense stellate hairs showed that elongate rostra were essential and the successful species had them (Carter 1982). Moran's (1986*b*) morphometric study of the aphid genus *Uroleucon* showed that trichome density was correlated with the length of the terminal rostral segment, and it is likely that the adaptation is associated with a similar problem.

The difficulty in penetrating a smooth hard surface has not been addressed specifically although there is a suggestion that many details of mouthpart structures relate to this, and the ability of the tarsi to obtain a firm purchase is also likely to be important. Edge feeding is a behavioural approach to gaining entry for biting insects, but species that feed from the leaf surface can be expected to have a wide gape and needle-like incisor teeth: characteristics described by L. Casher (unpublished results) for first instar caterpillars feeding on the surface of very hard, smooth leaves of evergreen oaks.

Grasshoppers feeding on hard cylindrical leaves and stems appear to have very enlarged paraglossae on the labium. Chapman (1966) described the mouthparts of one such species, *Xenocheila zarudnyi* which feeds on *Ephedra*, where the paraglossae are cone-shaped and extend back to the front legs. In discussing these and similar structures that have evolved independently in several different grasshopper subfamilies, Chapman quotes suggestions that enlarged paraglossae act like the jaws of a brace, gripping the tip of the cylinder while the mandibles cut through it.

3. LEAF HARDNESS OR TOUGHNESS AND GRASSHOPPER MANDIBLES

The majority of grasshoppers feed on many plant species, with herbaceous plants predominating, but a large proportion have become grass specialists. Indeed the evolution of grasses, perhaps in response to vertebrate grazing pressure, has had a profound influence on the evolution of grasshopper mandible morphology. This has been known for a long time and the differences between the mandibles of forb (nongrass) feeders and grass feeders have been described in detail (see, for example, Iseley 1944; Chapman 1964). Unlike some of the holometabolous insect larvae, grasshoppers have a clearly distinguished distal incisor area and a separate molar area. The two mandibles are asymmetrical with the left mandible closing over the right to bring the molar areas together. The inner surface of the left incisor region and the outer surface of the right incisor region are of particularly hard cuticle (Gardiner & Khan 1979). In forb feeders the incisor edge bears several sharp cusps or teeth, and the molar region is also strongly toothed when unworn. In grass

Table 1. *Examples of grasshopper species from different taxa which all share the characteristic graminivorous mandibles and are host specialists on the Poaceae*

(Different numbers represent independent evolutionary origins. From Chapman (1964, 1988) and Bernays & Chapman (1978).)

Romaleidae	*Epiprora hilaris*	1
Acrididae		
Hemiacridinae	*Laxabilla smaragdina*	2
Tropidopolinae	*Tristia marginicosta*	3
Oxyinae	*Oxya japonica*	4
Catantopinae	*Tristia conops*	5
Cyrtacanthacridinae	*Nomadacris septemfasciata*	6
Leptysminae	*Stenacris fissicauda*	7
Acridinae	*Acrida conica*	8
Oedipodinae	*Locusta migratoria*	8
Gomphocerinae	*Chorthippus curtipennis*	8
Truxalinae	*Mesopsis gracilicornis*	8

Table 2. *Relative leaf toughness or hardness in plants with different growth form, with leaves of herbaceous dicots standardized to 1*

(Partly after Bernays & Hamai (1987).)

Plant type	n^a	Relative toughness
palms: expanded fronds	8	9.8
woody plants: fully expanded leaves	89	6.3[b]
C4 grasses: all blades	34	6.2
C3 grasses: all blades	42	3.1
woody plants: new leaves	25	1.7[b]
herbaceous dicots: all leaves	166	1.0

[a] n, Number of species, each species figure based on 10–20 readings on apparently typical leaves.

[b] The ratio of old:new leaves of 3.7 is close to that of Raupp (1985) of approximately 4 for willow leaves.

specialists the incisor cusps tend to be chisel edged and the molar regions present ridges that appear to provide a grinding apparatus, reminiscent of the molar teeth in mammalian grazers (on a scale that is smaller by three orders of magnitude). Mixed feeders tend to have forbivorous mandibles, but of very stout construction. The interesting evolutionary factor here is the convergence of grass specialists on a common phenotype: the derivation of the graminivorous mandibles from the more generalized ones has probably occurred eight times (table 1).

Quantitative morphometric analyses were done by Patterson (1984) on overall dimensions. Forb feeders, in some subfamilies at least, have mandibles longer from hinge to distal cusps relative to the length of the hinge than do grass feeders, and the ratio of these values is clearly related to the proportion of forbs in the diet, at least in some groups. Maximum length of mandible, including incisor cusps, is achieved when 80% or more of the food is composed of herbaceous dicots. Grass specialists invariably have relatively short mandibles.

The benefits that accrue from the mandibular adaptations of graminivores are probably several. Firstly, the physical structure of grasses provides a difficult challenge for small herbivores in actually cutting through the many sclerophyllous parallel veins (Vincent 1982) which are responsible for the toughness (table 2), and the chisel-type incisor region is likely to be important for this task. Overall mandible shape differences will bring the occlusal surface nearest to the hinge in graminivorous forms, providing a mechanical advantage for the work needed as well as a stouter structure. The molar ridges are unlikely to be acting in precisely the same way as in grazing mammals owing to the differences in scale, and the fact that much ingestion of grasses in grasshoppers occurs without periods of grinding or chewing. However, a similar purpose may be served. It is possible that a single effective grinding action occurs when the piece of grass blade is grasped by the mandibles and passed back. Ab- and adaxial surfaces are sheared apart so that swallowed particles have a large exposure of parenchyma, potentially increasing the digestive area. Boys (1981) found that in a forb feeder 80% of ingested fragments had two epidermises, whereas in a grass feeder 80–100% of the fragments had only one epidermis. It should be noted that the grass specialist *Locusta migratoria* does pause and chew between successive bites, and that the chewing probably makes up about 10% of the meal time (Simpson *et al.* 1988). A very extreme case is the grasshopper *Microtylopteryx hebardi*. Its mandibles are similar to those of grass specialists and it feeds on extremely tough rainforest monocots, especially palms (see table 2): it has been observed to chew a cut particle of palm leaf up to twenty times before taking another bite (E. A. Bernays, unpublished results). Finally, the presence of silica in grasses may have had some influence on mandibular morphology: certainly the wear caused by feeding on grasses can be extreme (Chapman 1964), although perhaps the silica bodies in grass epidermis could be used in obtaining a purchase by the molar ridges.

In addition to form, it has been noted that grass-specialist grasshoppers have much larger mandibles and heads relative to body size compared with forb feeders, with mixed feeders being intermediate (Bernays 1986a). The log–log regressions of head mass against headless body mass have identical slopes but significantly different intercepts for the three types of feeders, with overall head sizes (as percent of headless body) being 23, 11 and 15% respectively (Bernays & Hamai 1987). Mandible and mandible adductor muscle mass together make up 49% of the total dry head mass with very little variation in relation to diet or to a variety of cryptic head shapes. The force needed to cut graminaceous tissue requires powerful muscles, and heads large enough to house them, and the conclusion must be that graminivores are well adapted for dealing with their food. The benefits probably relate both to improved handling and improved digestive processing, but the cost is large in terms of investment in head mass.

4. LEAF HARDNESS OR TOUGHNESS AND CATERPILLAR MANDIBLES

Unlike orthopterans, lepidopterous larvae have mandibles without a true molar area. In general the mandible is a simple structure with a row of simple incisor cusps. Various projections and teeth occur in different positions on the inner face, some of which are taxonomic characters of unknown function (Peterson 1962). Feeding habits are often not independent of taxonomy, adding a further difficulty in discerning adaptive value for particular morphological structures, but some contrasts can be made between species utilizing tough trees or grass leaves and those utilizing soft herbaceous plants. Unlike grasshoppers, a relatively small proportion have become strict graminivores,, whereas tree-feeding species are at least as abundant as herb feeders (Bernays & Barbehenn 1987).

The common observation that trees, shrubs and grasses have harder or tougher leaves than do herbaceous dicots has been quantified by Bernays & Hamai (1987) by using a penetrometer with a blade intended to represent the incisor edge of an insect mandible. Measured in arbitrary units, the ratios of toughness, with herbaceous plants as 1, are shown in table 2. Clearly the difficulty of biting through a leaf is dependent on several different factors, in particular the presence and arrangement of sclerophyllous tissues including lignified veins, and to a lesser extent, hard cuticles. Penetrometer measurements done with a pinpoint between veins allows independent measures of penetrability of upper and lower surfaces and L. Casher (unpublished results) found that in *Quercus agrifolia* the abaxial cuticle was harder than the adaxial.

Caterpillar species that have become grass specialists have, as in grasshoppers, relatively larger heads than those that feed on herbaceous dicots (Bernays 1986*a*). As grasses and other tough foliage is low in protein, the challenge is twofold.

1. An increased force must be developed, which, together with reduced contact area of mandible and leaf, will provide the necessary pressure (Wheater & Evans 1989). This requires adductor muscles with a large cross-sectional area.

2. More bites are needed to obtain enough food on the tough leaves, with the possibility that the increased work will cause an increase in muscle development.

The fact that greater positive allometric growth of the head occurred within a cohort of caterpillars of *Pseudaletia unipuncta* fed on tough grasses, compared with others on soft food, provides further evidence of the importance of head and mandible size.

All species examined that feed primarily on grasses and older tough leaves of woody plants have simplified mandibles in the later instars. Instead of the primitive row of sharp incisor cusps along the edges that loosely overlap (left over right and vice versa), there is a smooth incisor edge, usually devoid of cusps, and often with a groove or ridge of some kind into which the opposing incisor region fits. Instead of a cutting and tearing behaviour, the action is more akin to scissor or tin-snip action. Examples in table 3 show the convergence of this trait in several different families. Godfrey *et al.* (1989) examined 143 species of Notodontidae and the few species having incisor teeth are those specializing on very soft new leaves. Similarly, Bernays & Janzen (1988) examined 12 species of Saturniidae, all of which feed generally on mature tree leaves, and all had the incisorless mandible type, albeit each with characteristic details. Barbehenn (1989) found incisorless mandibles characteristic of 18 species of Hesperiidae that feed on grasses or trees. As with grasshoppers, the ratio of length of mandible to length of hinge is lowest for these tree-feeders and grass-feeders, resulting in a very stout structure. The convergence of the phenological characters is clearly indicative of an adaptive function, apparently related to the toughness or hardness of the foliage consumed.

Several authors have noted that many of the species with toothless snipping mandibles in late instars have well-developed teeth in the earlier instars. They feed between major veins and their problem is partly dealing with smaller veins, but more often with a hard cuticle. Indeed the teeth seem to be used in gouging out mesophyll between the veins, leaving the rest of the leaf intact. The change in morphology and the switch to a feeding habit in which whole leaf thickness is cut occurs at different instars depending on species (Hagen & Chabot 1986; Godfrey *et al.* 1989; L. Casher, unpublished results). On the other hand, some very small species of caterpillar that feed on tough tree leaves

Table 3. *Examples of species pairs within families where one has typical lepidopteran incisor teeth and the other has adopted the toothless snipping mandibles associated with their feeding on tough leaves of trees or grasses*

	food: new leaves or herbaceous	food: tree leaves or grasses	
family	Incisor teeth	No incisor teeth	reference
Notodontidae	*Crinodes besckiae*	*Cerura vinula*	1
Dioptidae	*Cyanotricha necyria*	*Phryganidia californica*	1, 2
Noctuidae	*Spodoptera littoralis*	*Spodoptera exempta*	3
Sphingidae	*Manduca rustica*	*Enyo ocypete*	4
Hesperiidae	*Pyrgus communis*	*Erynnis tristis*	5
Pyralidae	*Uresiphita reversalis*	*Chilo partellus*	6

[a] 1, Godfrey *et al.* (1989); 2, L. Casher (unpublished results); 3, Brown & Dewhurst (1975); 4, Bernays & Janzen (1988); 5, Barbehenn (1989); 6, E. A. Bernays (unpublished results).

remain skeletonizers and retain strongly toothed mandibles (L. Casher, unpublished results). The relation between leaf type, feeding behaviour and gross mandible morphology is very compelling, and in addition there is a suggestion that the snipping mandible is associated with toughness (fibres, sclerenchema, lignin) and very sharp teeth with cuticle hardness.

The cutting or snipping action of mandibles dealing with tough leaves results in rather standard ingested particle size for a particular size of mandible, which was considered a constraint on size by Bernays & Janzen (1988) because large particles might be inefficiently digested. Although there may be upper limits to ideal bite sizes for digestion in caterpillars, Barbehenn (1989) found that sizes were not correlated with digestibility levels, even though at larger bite sizes as few as 6 % of leaf cells were physically damaged. In spite of the commonly held belief that folivores chew their food, most caterpillars (unlike grasshoppers) ingest unmacerated leaf tissue and can efficiently digest the nutrients therein (Harvey 1975; Hocking & Depner 1961; Rybicki 1957; Barbehenn 1989). This is achieved through a combination of high gut pH (Jones *et al.* 1989), presence of surfactants (Martin & Martin 1984) and efficient phospholipases that rapidly degrade cell membranes allowing rapid diffusion through the plasmodesmata, which have diameters of over 60 nm after disruption of the desmotubule (Barbehenn 1989). Furthermore, Barbehenn found that in two grass-feeding species, protein was quite efficiently removed even from C4 grasses where about half of the leaf protein is contained within the bundle sheath cells the walls of which are highly lignified. In fact, digestibility was equally good in the hesperiid with a food transit time of 8–20 h and noctuid caterpillar with a transit time of 2–4 h.

From this it would appear that incisor teeth, even though they result in leaf particles with relatively large circumference to area ratios (Bernays & Janzen 1988), have no role in assisting digestive processing by caterpillars, especially as species with and species without all have similar digestibilities of standard seedling wheat (Bernays & Barbehenn 1987).

Convergence can also be seen in mandibles of large caterpillar species that feed on soft flaccid leaves. There is considerable development of projections or teeth, not just on the edge of an incisor area but on the whole distal region and inner face of the mandibles; teeth of one mandible fit into depressions on the opposing mandible. Dorsally, where the two mandibles first meet during a bite, there is a serrate edge apparently for penetration of the leaf. This pattern, with many variants, can be seen in Sphingidae (Bernays & Janzen 1988) and in certain examples elsewhere (Godfrey *et al.* 1989). Interestingly, in the sphingids, at least one species, *Enyo ocypete*, has returned to the toothless snipping type and is a specialist on the very tough leaved *Tetracera volubilis* (table 3).

Bernays & Janzen (1988) contrasted the snipping mandibles of Saturniidae with the complex interlocking mandibles of Sphingidae and concluded that differences related to both handling and digestibility.

In the light of Barbehenn's (1989) findings the emphasis should perhaps be on efficiency of handling. Tough leaves are efficiently handled by the snipping, scissor action, whereas the softer more flaccid leaves are more efficiently ingested by the tearing, crushing action. Because both leaf types can be eaten by both caterpillar types, the concept of efficient handling becomes more subtle: rate of food intake may be under intense selection pressure if feeding increases the chances of predation. Certainly tough leaves take longer to ingest by caterpillars not specifically adapted than by those that are (Bernays 1986*a*; Devitt & Smith 1985), whereas feeding caterpillars are about 30 % more vulnerable to predation by vespid wasps than are resting ones (E. A. Bernays, unpublished results), so that reduction in time spent feeding could provide a considerable selective advantage. In addition, the studies of Heinrich (1983) suggest that the act of feeding by caterpillars increases vulnerability to bird predation.

Wear is a potential problem for mandibulate insects feeding on hard, tough or abrasive material, and it is known to have a large impact on the form of grasshopper (Chapman 1964) and beetle (Raupp 1985) mandibles, increasing the time taken to eat a given quantity of leaf. Wear also occurs in caterpillar mandibles and measurements were made of the incisor:hinge ratio in newly molted fifth instar saturniid *Rothschildea lebeau* or on the exuviae. Big changes occur – a 10 % reduction in incisor length – but interestingly the edge appears to be more or less maintained, indicating that the mandibles may be self sharpening, perhaps by having a softer cuticle on the inner face than on the outer face. This ability may be a feature of species feeding on tough leaves, and it has been noticed that the weakly toothed mandibles of third instar *Phryganidia californica* are quickly worn down to the smooth type (unpublished observations). Photographs published by Djamin & Pathak (1967) suggest that the stem borer, *Chilo suppressalis*, maintains toothed mandibles on low silica hosts but these become toothless snipping structures on high silica rice.

5. CHEMORECEPTORS

Number and arrangement of chemoreceptors is apparently related to host-specific behavioural adaptation in some groups. In the aphid genus *Uroleucon*, alatae of species on hosts that have a clumped distribution have significantly less extensive rhinaria on the antennae than those that utilize highly dispersed plant species (Moran 1986*b*), suggesting that host location needs will determine this aspect of aphid morphology. This idea is supported by the differences found in relation to aphid polyphenism: less mobility, as judged by lack of wings and shorter legs, is always associated with fewer antennal chemoreceptors.

Chapman (1982) undertook an extensive study of sensilla and found strong evidence of larger numbers being associated with broader host ranges. At higher taxonomic levels it is possible, as he suggests, that smaller numbers permit or promote narrow host range:

diets are markedly broader in the orthopteroid orders where sensillum numbers are extremely large. On the other hand, among grasshoppers, where specialized diets are clearly derived, it appears that narrow diet has led to a reduction in chemoreceptors. The subfamilies that are characteristically grass feeding all have fewer receptors, and there is evidence of considerable further reduction in all the specialists examined (Chapman & Thomas 1978). For example in the Gomphocerinae, where most species are graminivorous, the polyphagous species examined have larger numbers, whereas the extreme specialists have fewer (Chapman & Fraser 1989). Chapman (1982) suggests that the functional significance of the morphological specialization perhaps relates to the relative ease of determining appropriate food among specialists, because of labelled chemosensory lines for host-specific chemicals. However, information so far suggests that grass specialists select food plants on the basis of absence of deterrents instead of via grass-specific cues (Bernays & Chapman 1978), so the explanation may not be so simple.

6. INTERNAL MORPHOLOGY

There are few studies on what morphologies are under selection pressure internally. Chapman (1988) examined the guts of 170 species of grasshopper, in which there are six gastric caeca that typically have anterior and posterior arms. In grass feeders, however, the posterior arms were invariably reduced or absent. The average ratio of posterior arm length to anterior arm length was twice as high in forb feeders as in graminivores, and he suggested that this morphological difference may relate to reduced needs for detoxification of plant secondary metabolites in grass feeders.

In the examination of fifteen grasshopper species the foregut volume was also found to be significantly greater in polyphagous species than in graminivores (E. A. Bernays & Raubenheimer, unpublished results). The most extreme polyphagous forb feeder, *Taeniopoda eques*, had a full crop mass that was 50% of the total body mass, whereas at the other extreme in four of the grass specialists it was about 10%. Whether this relates to chemical composition is unknown, but other possibilities exist. From the few observations made so far, one could suggest that the grass feeders and other relative specialists are less exploratory than the polyphages and perhaps less likely to become separated from highly suitable food resources.

Plant material being commonly low in protein, insects could possibly conserve on protein by reducing investment in cuticle. This could be seen as one of the benefits accruing to soft bodied holometabolous larvae of caterpillars having a thin cuticle, in comparison with grasshoppers, in which nearly half of the dry body mass is proteinaceous cuticle (Bernays 1986b). Rees (1986) used similar reasoning when he found that leaf-feeding beetles and bugs had thinner cuticles than carnivorous ones.

Examination of internal organs will certainly reveal more: there are likely to be variations in relation to specific host chemistry, water content, symbiotes and lifestyle. Large structural differences exist at the ordinary level and there are well known adaptations for regulation of water content: it would be surprising if there were not an infinite variety of adaptations between and within species.

7. EVOLUTION AND CONCLUSIONS

The evolution of morphological adaptations is largely veiled in the past, and direct measures of any evolutionary change are rare. Natural rates of evolution of morphological traits in insects are thus virtually unknown. In a recent study Carroll & Boyd (1991) provide direct evidence of evolution by natural selection of beak length in *Jadera haematoloma* bugs over 20–50 years. This insect feeds on seeds contained within the fruits of various species of Sapindaceae, and its stylets must reach through the fruit to the seeds. In the southern U.S.A., recently introduced plants have been colonized by this bug which has formed host races with beak lengths either longer or shorter to suit large or small fruits. Measures of museum specimens have provided the evidence that ancestral populations were different and had the requisite variation to respond to novel selection pressures seen over decades. Directional selection for beak length has been uniformly rapid in three different host races; differences among the races in patterns of phenotypic change also imply differences in the genetic processes.

The examples of morphological adaptations given in this paper show quantitative changes that are probably achieved by allometric growth. In addition, the ontogenetic changes may partly reflect effects of experience, as in the case of hard foods that cause greater positive allometry in *Pseudaletia unipuncta* larvae. The development of snipping lepidopterous mandibles in species that start off with toothed incisor regions may also be influenced by the feeding regime, directly by wear, and perhaps indirectly by the effect of wear on subsequent patterns of cuticle deposition.

Adaptive morphology of insect structures related to host use is present wherever it is looked for. Because morphological characters seem to respond rapidly to selection, they may be limited in their use for inferring phylogenies. In addition, care is needed to interpret correlations found in relation to host characters, as these may arise through having clusters of related similar species. Thus phylogeny and host-related morphological adaptation need to be studied together (Felsenstein 1985). Finally, morphology is inextricably associated with behaviour and ecology, and in studying the adaptations of insect herbivores with plants, particular account must be taken of predators, and perhaps other mortality factors, that influence evolution of insects in relation to their host plants.

REFERENCES

Barbehenn, R. V. 1989 The nutritional ecology and mechanisms of digestion of C3 and C4 grass-feeding

Lepidoptera. Ph.D. thesis, University of California at Berkeley.

Bernays, E. A. 1986*a* Diet-induced head allometry among foliage-chewing insects and its importance for graminivores. *Science, Wash.* **231**, 495–497.

Bernays, E. A. 1986*b* Evolutionary contrasts in insects: nutritional advantages of holometabolous development. *Physiol. Ent.* **11**, 377–382.

Bernays, E. A. 1989 Host range in phytophagous insects: the potential role of generalist predators. *Evol. Ecol.* **3**, 299–311.

Bernays, E. A. & Barbehenn, R. V. 1987 Nutritional ecology of grass foliage-chewing insects. In *Nutritional ecology of insects, mites, spiders, and related invertebrates* (ed. F. Slansky & J. G. Rodriguez), pp. 147–175. New York: Wiley & Sons.

Bernays, E. A. & Chapman, R. F. 1978 Plant chemistry and acridoid feeding behaviour. In *Biochemical aspects of plant and animal coevolution* (ed. J. B. Harborne), pp. 99–141. London: Academic Press.

Bernays, E. A. & Hamai, J. 1987 Head size and shape in relation to grass feeding Acridoidea (Orthoptera). *Int. J. Morphol. Embryol.* **16**, 323–36.

Bernays, E. A. & Janzen, D. H. 1988 Saturniid and sphingid caterpillars: two ways to eat leaves. *Ecology* **69**, 1153–1160.

Boys, H. 1981 Food selection by some graminivorous Acrididae. D. Phil. thesis, Oxford University, U.K.

Brown, E. S. & Dewhurst, C. F. 1975 The genus *Spodoptera* (Lepidoptera, Noctuidae) in Africa and the Near East. *Bull. Ent. Res.* **65**, 221–262.

Carroll, S. P. and Boyd, C. 1991 Host race radiation in the soapberry bug: introduction history of the host plants and the evolution of beak length. *Evolution.* (In the press.)

Carter, C. I. 1982 Susceptibility of *Tilia* species to the aphid *Eucallipterus tilae.* In *Proc. 5th Int. Symp. Insect–Plant Relat.* (ed. J. H. Visser & A. K. Minks), pp. 421–423. Wageningen: Pudoc.

Chapman, R. F. 1964 The structure and wear of mandibles in some African grasshoppers. *Proc. zool. Soc. Lond.* **142**, 107–121.

Chapman, R. F. 1966 The mouthparts of *Xenocheila zarudnyi* (Orthoptera, Acrididae) *J. Zool.* **148**, 277–288.

Chapman, R. F. 1982 Chemoreception: the significance of sensillum numbers. *Adv. Insect Physiol.* **16**, 247–356.

Chapman, R. F. 1988 The relationship between diet and the size of the midgut caeca in grasshoppers (Insecta: Orthoptera: Acridoidea). *Zool. J. Linn. Soc.* **94**, 319–338.

Chapman, R. F. & Fraser, J. 1989 The chemoreceptor system of the monophagous grasshopper, *Bootettix argentatus* Bruner (Orthoptera: Acrididae). *Int. J. Insect Morphol. Embryol.* **18**, 111–118.

Chapman, R. F. & Thomas, J. G. 1978 The numbers and distribution of sensilla on the mouthparts of Acridoidea. *Acrida* **7**, 115–148.

Devitt, B. D. & Smith, J. J. B. 1985 Action of mouthparts during feeding in the dark-sided cutworm, *Euxoa messoria* (Lepidoptera: Noctuidae). *Can. Ent.* **117**, 343–349.

Djamin, A. & Pathak, M. D. 1967 Role of silica in resistance to Asiatic rice borer, *Chilo suppressalis* (Walker), in rice varieties. *J. econ. Ent.* **60**, 347–351.

Eastop, V. F. 1985 The acquisition and processing of taxonomic data. In *Proc. Int. Aphidol. Symp. Jablonna 1981* (ed. H. Szelegiewicz), pp. 245–270. Wroclaw, Poland: Ossolineum.

Felsenstein, J. 1985 Phylogenies and the comparative method. *Am. Nat.* **125**, 1–15.

Gardiner, B. G. & Khan, M. F. 1979 A new form of cuticle. *Zool. J. Linn. Soc.* **65**, 91–94.

Gillette, C. P. 1911 A new genus and four new species of Aphididae. *Ent. News* **22**, 440–444.

Godfrey, G. L., Miller, J. S. & Carter, D. J. 1989 Two mouthpart modifications in larval Notodontidae (Lepidoptera): their taxonomic distributions and putative functions. *J. New York Ent. Soc.* **97**, 455–470.

Hagen, R. H. & Chabot, J. F. 1986 Leaf anatomy of maples (*Acer*) and host use by Lepidoptera larvae. *Oikos* **47**, 335–345.

Harvey, G. T. 1975 Nutritional studies of eastern spruce budworm (Lepidoptera: Tortricidae) II. Starches. *Can. Ent.* **107**, 717–728.

Heinrich, B. 1983 Caterpillar leaf damage, and the game of hide-and-seek with birds. *Ecology* **64**, 592–692.

Hocking, B. & Depner, K. R. 1961 Larval nutrition in *Agrotis orthogonia* (Lepidoptera: Phalaenidae): digestive enzymes. *Ann. ent. Soc. Am.* **54**, 86–98.

Hoffman, G. D. and McEvoy, P. B. 1985 The mechanisms of trichome resistance in *Anaphalis margaritacea* to the meadow spittlebug *Philaenus spumarius. Entomol. Exp. Appl.* **39**, 123–129.

Iseley, F. B. 1944 Correlation between mandibular morphology and food specificity in grasshoppers. *Ann. entomol. Soc. Am.* **37**, 47–67.

Jones, C. G., Hare, J. D. & Compton, S. J. 1989 Measuring plant protein with the Bradford assay. *J. chem. Ecol.* **15**, 979–992.

Kennedy, C. E. J. 1986 Attachment may be a basis for specialization in oak aphids. *Ecol. Ent.* **11**, 291–300.

Lee, Y. I., Kogan, M. & Larsen, J. R. 1986 Attachment of the potato leafhopper to soybean plant surfaces as affected by morphology of the pretarsus. *Entomol. Exp. Appl.* **42**, 101–108.

Martin, M. M. & Martin, J. S. 1984 Surfactants: their role in preventing the precipitation of protein by tannins in insect guts. *Oecologia, Berl.* **61**, 342–345.

Moran, N. A. 1986*a* Benefits of host plant specificity in *Uroleucon* (Homoptera: Aphididae). *Ecology* **67**, 108–115.

Moran, N. A. 1986*b* Morphological adaptation to host plants in *Uroleucon* (Homoptera: Aphididae) *Evolution* **40**, 1044–1050.

Patterson, B. D. 1984 Correlation between mandibular morphology and specific diet of some desert grassland Acrididae (Orthoptera). *Am. Midl. Nat.* **111**, 296–303.

Peterson, A. 1962 *Larvae of insects. Part 1, Lepidoptera and plant-infesting Hymenoptera.* (315 pages.) Columbus, Ohio.

Raupp, M. J. 1985 Effects of leaf toughness on mandibular wear of the leaf beetle *Plagiodera versicolora. Ecol. Ent.* **10**, 73–79.

Rees, C. J. C. 1986 Skeletal economy in certain herbivorous beetles as an adaptation to poor dietary supply of nitrogen. *Ecol. Ent.* **11**, 221–228.

Rybicki, M. 1957 Mechanism of digestion of leaves of green plants by some lepidopterous caterpillars. *Acta biol. exp., Warsaw* **17**, 289–316.

Stork, N. 1980 Experimental analysis of adhesion of *Chrysolina polita* (Chrysomelidae: Coleoptera) on a variety of surfaces. *J. exp. Biol.* **88**, 91–107.

Simpson, S. J., Simmonds, M. S. J., Wheatley, A. R. & Bernays, E. A. 1988 The control of meal termination in the locust. *Anim. Behav.* **36**, 1216–1227.

Southwood, T. R. E. 1973 The insect–plant relationship – an evolutionary perspective. *Symp. R. ent. Soc. Lond.* **6**, 3–30.

Southwood, T. R. E. 1986 Plant surfaces and insects – an overview. In *Insects and the plant surface* (ed. B. Juniper & T. R. E. Southwood), pp. 1–22. London: Edward Arnold.

Vincent, J. F. V. 1982 The mechanical design of grasses. *J. Mater. Sci.* **17**, 856–860.

Wheater, C. P. & Evans, M. E. G. 1989 The mandibular forces and pressures of some predaceous Coleoptera. *J. Insect Physiol.* **35**, 815–820.

Winder, L. 1990 Predation of the cereal aphid *Sitobion avenae* by polyphagous predators on the ground. *Ecol. Ent.* **15**, 105–110.

Discussion

E. A. JARZEMBOWSKI (*Booth Museum of Natural History, Brighton, U.K.*). There has been much discussion recently on the functional morphology of flight in insects. *Rhyniella praecursor*, the earliest known insect (hexapod) was a flightless apterygote and lived in an arachnid-infested bog as described by Professor Chaloner (this symposium).

By analogy with living insects (e.g. *Podura*), would not the long pretarsi of this small wingless insect have enabled it to walk on the surface of water and thus avoid predation?

E. A. BERNAYS. This is an interesting idea but maybe walking on the surface had more importance for prevention of drowning!

S. B. MALCOLM (*Department of Biology, Imperial College, Silwood Park, U.K.*). Professor Bernays suggests that natural enemies constrain the movement of cryptic insect herbivores and select indirectly for increased morphological feeding adaptations so that plant food can be ingested as effectively as possible. Do you think that aposematic insect herbivores are less constrained by the need for morphological feeding adaptations because they forage conspicuously? Perhaps aposematic generalists show less morphological feeding adaptation than aposematic specialists and both are less specialized morphologically than cryptic feeders.

E. A. BERNAYS. This would be a most interesting approach to testing my hypothesis. I cannot think of any studies to draw upon at this point but it is well worth investigation. Of course even aposematic species have their enemies so the pattern is probably not clear cut.

Herbivory and the evolution of leaf size and shape

V. K. BROWN and J. H. LAWTON

Department of Biology and NERC Centre for Population Biology, Imperial College, Silwood Park, Ascot SL5 7PY, U.K.

SUMMARY

Why do leaves have such varied sizes and shapes? Part of the answer lies in physiological and biomechanical demands imposed by different habitats; selective forces that are now reasonably well understood. In contrast, the impact of herbivores on the evolution of leaf size and shape has rarely been investigated and is poorly understood. There are at least six ways in which herbivores, particularly vertebrates and insects, may have influenced the evolution of leaf size and shape, favouring leaf morphologies that differ from those dictated by physiological and biomechanical constraints acting on plants. They are mimicry, not only of leaves of other plant species but also grazed leaves and inanimate objects; crypsis; physical barriers to being eaten; interspecific differences in leaf morphology to reduce recognition by herbivores; very small or highly divided and dissected leaves that reduce feeding efficiency; and different adult and juvenile foliages. There is an urgent need for studies specifically designed to investigate the impact of herbivores on leaf size and shape.

1. INTRODUCTION

Why are there so many sizes and shapes of leaves? Within inevitable phylogenetic and developmental constraints, part of the answer lies in the physiological and biomechanical demands imposed by different habitats (i.e. light regimes, temperatures, humidities and wind speeds) interacting with nutrient and water availability (Böcher & Lewis 1962; Horn 1971; Parkhurst & Loucks 1972; Givnish 1979, 1986; Chabot & Hicks 1982). Models that assume that natural selection minimizes the costs of supporting leaves and maximizes whole-plant carbon gain predict much of the variation observed in the sizes and shapes of real leaves (Givnish 1986 and references therein).

The purpose of this paper is not to deny the importance of physiological and biomechanical constraints on the design of leaves, but rather to draw attention to the possibility that herbivory is an additional but neglected constraint. Herbivory as a selective force on the economy of leaf design is mentioned only once, and then only in passing, in the seven hundred or so pages of Givnish (1986, p. 245). Our thesis is that leaf size and shape influence the impact of herbivores, and are therefore open to selection by herbivores. It follows that some of the variation in the design of leaves seen within and between habitats may be an evolved response by plants to herbivory, rather than to light, temperature, wind and so on. How much herbivores can change the design of leaves away from purely physiological and biomechanical optima is an open question, but the answer is unlikely to be 'not at all'.

The paper is not about leaf toughness, or microscopic features (hairs, trichomes, etc.), or secondary chemistry, all of which have been implicated as plant defences against herbivory (see, for example, Southwood (1987)). It is concerned solely with the sizes and shapes of leaves. However, size and shape may covary with other characteristics (see, for example, Givnish (1979)). Consequently, it may be difficult to attribute herbivore responses primarily, or mainly, to leaf size and shape, a caveat that must be continually borne in mind.

First, we examine how plants in New Zealand responded to a world without native, mammalian herbivores, but subject to grazing by giant extinct birds, the moas. We then consider the impact of mammalian herbivores on the evolution of leaf size and shape. The remainder of the paper focuses on insects, being the group with which we are personally most familiar. Much of the evidence assembled for the role of herbivores, both vertebrate and invertebrate, in the evolution of leaves is anecdotal and imperfect. Our intention is to stimulate more thought and more work, rather than to provide definitive answers.

There are at least six ways in which evolutionary responses by plants to herbivory may have influenced the sizes and shapes of leaves. They are: (i) mimicry, particularly of other species of plants that are unsuitable as food, but also of herbivore-damaged leaves signalling, for example, induced chemical defences; (ii) a set of leaf characteristics that range from crypsis to mimicry of inanimate objects; (iii) leaves modified to provide direct physical defences in the form of spines, or other physical barriers; (iv) divergence in leaf size and shape between species in the same habitat, a phenomenon that may reduce the risk of a particular plant species being recognized as potential food by visually searching herbivores; (v) very small or highly divided leaves that reduce the foraging efficiency of herbivores, or make effective feeding impossible; (vi) different adult and juvenile foliage in the some species, a tactic that may have evolved because the risk of herbivore attack changes with the maturity of the plant.

2. MOAS AND THE EVOLUTION OF THE NEW ZEALAND FLORA

The flora of New Zealand evolved in the absence of mammalian herbivores, but subject to grazing by moas. At least twelve species of moas, ranging in size from 20 to 200 kg and approaching 3 m tall, are known from their remains. All were exterminated by the Maoris, who first colonized New Zealand in the eleventh century. Relations between moas and plants are reconstructed and reviewed by Atkinson & Greenwood (1989).

Browsing and grazing mammals detect food by smell, touch and sight, but lack colour vision; they cut chunks of plant tissue with a rotatory action of the jaw. Moa browsing and grazing subject plants to very different selection pressures. Moas, in common with extant, diurnal birds presumably had acute colour vision, but a poor sense of smell. Birds peck, and can only bite food with a scissor action, swallowing it without chewing. Atkinson & Greenwood argue that four widespread features of the New Zealand flora evolved in response to moa grazing. They are spiny tussocks, mimicry, reduced visual apparency and divarication.

Spines are uncommon among New Zealand plants. However, the genus *Aciphylla* (Apiaceae) is exceptional, and several species form vicious, hemispherical tussocks, presenting a surface of closely spaced, rigid, sharp points borne on modified leaves. The hypothesis, first mooted by Wallace in 1889, is that the unusual leaves and growth form of aciphyllas may have reduced attack by moas, seeking to feed on the nutritious leaf bases. A bird pecking into the tussocks would risk impaling an eye, unless it was extremely careful. Consistent with this hypothesis, *Aciphylla* species growing in habitats where moas were rare or absent have reduced, or no leaf spines. For example, the Australian *A. glacialis* has short, fern-like leaves. Moreover, mature, spiny aciphyllas have seedlings with flaccid, grass-like leaves but, as plants enlarge to tussocks and become clearly visible, the juvenile foliage is quickly replaced.

By relying on acute vision for food selection, moas may have favoured the evolution of mimicry and reduced visual apparency in potential food plants. Atkinson & Greenwood recognize two sorts of models for mimetic species, namely other, unpalatable plants and dead twigs. For example, *Celmisia lyalli* and *C. petriei* (Compositae) look rather like spiny aciphyllas. Several species of woody plants mimic unpalatable species when small, but have different, non-mimetic leaves when they are larger and hence above the reach of browsing moas. Mimicry of dead twigs is illustrated by the forest vine, *Parsonia capsularis* (Apocynaceae), which has very narrow, brown leaves blotched with grey when young, and more normal, green leaves with increasing height above the ground. It is unclear from Atkinson & Greenwood's account whether mature leaves are also a different shape. Leaves of other species certainly change shape as the plant matures. *Pittosporum patulum* (Pittosporaceae), a small tree, has deep purple-bronze juvenile leaves that are long and very narrow (2–4 mm wide) with toothed edges; adult leaves, presumably above the reach of browsers, are a completely different shape, untoothed, 10–15 mm wide and green. Juveniles of *P. patulum* do not mimic twigs, but are very difficult to see on the forest floor. It is representative of a number of New Zealand forest plants with cryptic, often brown juvenile foliage differing markedly in size or shape from the green foliage of adults; some of these species appear to be twig mimics, others are just very cryptic (J. H. Lawton, personal observation).

Divaricating shrubs are also a common feature of the New Zealand flora. The divaricating (wide branching angle) growth form has evolved several times, creating shrubs with interlacing branches, reminiscent of bundles of lightly crumpled and folded chicken wire. All have small leaves, sparse or absent on the exterior of the plant, but more dense towards the centre of the shrub. Atkinson & Greenwood suggest that the mechanical restriction presented by interlacing branches and small, inaccessible leaves would have reduced the bite size and biting rate of moas, and hence may have evolved in response to moa browsing, because they make feeding uneconomic. Other features of divaricating plants (for instance, springy, tough twigs that are difficult to pull off) may also have conferred protection by reducing moa feeding rates. *Hoheria angustifolia* (Malvaceae) is representative of a group of nine species that are divaricating when small, but which transform into normal, small trees with larger leaves above about 3 m, that is, just out of reach of the larger moas.

To the eyes of a European biologist, all these features of the New Zealand flora are peculiar, particularly the widespread prevalence of divaricating shrubs, and plants with leaves that change markedly in size, shape and colour as the plant grows. The hypothesis that moas may have driven the evolution of this suite of characteristics is simple and elegant. It may also be wrong. No other, single hypothesis satisfactorily unites these observations, but alternative hypotheses for particular features and phenomena exist, and not all the evidence adduced by Atkinson & Greenwood (1989) is beyond criticism (see, for example, McGlone & Webb (1981); McGlone (1991); M. McGlone (personal communication). Space prevents a full discussion, but a case can be made, for instance, that the divaricating growth form is advantageous in dry, windy conditions that prevailed over large parts on New Zealand until comparatively recently. That divaricating plants now occur outside such habitats may have as much to do with phylogenetic constraints and the pool of potential colonists available to occupy more mesic habitats as it has to do with moas.

3. VERTEBRATE GRAZERS IN THE REST OF THE WORLD

(a) Spinescent leaves and other changes in leaf morphology

So much of the world's flora has evolved in the presence of browsing and grazing mammals that we may fail to recognize their pervasive influence on leaf design. Spines and thorns certainly play a role in defence against browsing mammals, and sometimes, as in holly (*Ilex* spp.), they adorn the edges of leaves (where they may also serve to deter insect herbivores; see Ehrlich & Raven 1967). However, leaves with spines large enough to deter mammals are not common. In the flora of southern Africa, spinescent plants make up between 3 and 12% of species, depending upon the biome; of these, species with spiny leaf-tips and margins are a minority, constituting between 1 and 2% of the total flora and 9–52% of spinescent species (Milton 1991). In detailed line transects, Milton reported 12 species with spiny leaves, from a sample of 47 species of spinescent plants. It would be interesting to know whether leaf spines are comparatively rare because they are less efficient physical defences than tougher thorns on stems and branches; or because it is more costly to manufacture short-lived spines on leaves that will eventually be shed; or because spines compromise the physiological and biomechanical properties of leaves (see Givnish (1979) for further discussion).

Spines aside, certain leaf shapes may reduce the impact of grazing or browsing mammals, by allowing a partly eaten leaf to continue to function. For example, if leaflets are bitten off pinnate leaves, the remaining intact leaflets may be less prone to premature shedding or to the entry of pathogens through the wound than a simple leaf of the same total area, subject to the same percentage removal of tissue. Very small, pinnate leaves may also make browsing by mammals less economic (M. Coe, personal communication). We know of no tests of these possibilities.

(b) *Crypsis and mimicry*

Crypsis and mimicry could also evolve in response to mammalian herbivores, although examples that have received even cursory examination are hard to find, with the possible exception of Australian mistletoes (discussed below). There is evidence, however, that some weeds mimic the crops in which they grow, the selective pressure being hand weeding (Barrett 1983): a form of artificial selection that resembles heavy grazing by a visually searching herbivore. Furthermore, it is difficult to think of a good reason, other than escape from grazing mammals, why the paired, fused and fleshy leaves of about fifty species of South African stone plants (*Lithops* spp.) (Aizoaceae) should so beautifully mimic the stones and pebbles among which they grow, although Everard & Morley (1970) describe the idea as controversial. Barlow & Wiens (1977) briefly review mimicry in *Lithops* and in other, unrelated, succulents in several families. In Europe, the foliage and general appearance of vegetative plants of white dead-nettles (*Lamium album*) (Labiatae) bear a close resemblance to stinging nettles (*Urtica dioica*) (Urticaceae). Stinging hairs deter soft-muzzled, grazing mammals, suggesting that dead-nettles are harmless Batesian mimics. However, many other labiates that do not closely mimic nettles have ovate leaves with serrate margins, so if this is a case of true mimicry, it may have involved rather little modification in leaf shape.

(c) *Australian mistletoes that mimic their host-plants*

Several dominant genera of Australian trees (e.g. *Eucalyptus*, *Acacia*, *Casuarina*) have mistletoes (Loranthaceae) with leaves that resemble those of the host (Barlow & Wiens 1977). From a total of 64 Australian mistletoe species, 20 are highly host-specific mimics, and a further 16 species are mimics of one or more of their several hosts. Leaves of a wide variety of sizes and shapes are involved: filiform, falcate, lanceolate, elliptical and even the rounded-oval, amplexicaul juvenile leaves of *Eucalyptus*. Only evergreen hosts serve as models; mistletoes do not mimic deciduous hosts on which they would obviously be very conspicuous after leaf fall (J. L. Harper, personal communication).

The convergence in appearance between hosts and parasites appears not to be due to selection imposed by similar physical environments. Rather, Barlow and Wiens argue, it is most easily explained as protective mimicry against grazing by arboreal, herbivorous marsupials, particularly ringtails (*Pseudocheirus*) and possums (*Trichosurus*) in the extant fauna, although it may have evolved in the face of grazing pressures from other genera, now extinct. They argue that possums eat mistletoes and have excellent vision; they also provide anecdotal evidence to suggest that possums can have a considerable impact on the plants. Moreover, mimicry is most prevalent in species of mistletoes with high levels of foliage nitrogen, perhaps because they are most sought after by mammalian herbivores (Ehleringer *et al.* 1986).

However, not all biologists with a good knowledge of Australian mistletoes are agreed about the critical facts of Barlow and Wiens's arguments, still less about their interpretation (Atsatt 1983). Despite mimetic leaves, some mistletoes, Atsatt argues, are conspicuous because they stand out as dense clumps among sparse host foliage, or because they match the shape but not the colour of host leaves, and because mistletoes bear obvious flowers and berries. More startling, in limited feeding trials with a single possum, mistletoes were always rejected in favour of host foliage, except for 'small, exploratory bites upon the first encounter'. *Ergo*, mistletoes are unpalatable, an inference confirmed by extensive field searches that revealed little evidence of possum grazing on the plants (although it is clearly impossible to rule out selection by other arboreal herbivores that are now extinct). Finally, mimetic mistletoes are not confined to Australia; cases are reported from Africa and India, as well as North, Central and South America.

We are struck by the fact that both Atsatt (1983) and Barlow & Wiens (1977) dismiss an alternative hypothesis for the evolution of foliage mimicry in mistletoes. In Australia, the most damaging mistletoe herbivores are not vertebrates, but the caterpillars of lycaenid and pierid butterflies. The interaction between butterflies and mistletoes is complicated, at least for the lycaenids, by a mutualistic association between caterpillars and protective ants (Atsatt 1981), and Atsatt argues that there is no evidence that mistletoe leaf shape is important in host-plant recognition or selection by adult butterflies. But as far as we are aware nobody has tested adult, mistletoe-associated butterflies, under properly controlled conditions, to see if they respond to leaf shape. Females of at least one Australian lycaenid (*Jalmenus evagoras*), albeit not a mistletoe feeder, appear to use vision when searching for hosts (D. Nash, personal communication).

It is to the role of leaf shape in another butterfly–plant interaction that we now turn.

4. HELICONIUS BUTTERFLIES AND PASSIFLORA VINES

In the tropical rain forests and secondary woodlands of Central and South America, some of the most conspicuous insects are brightly coloured *Heliconius* butterflies. Their caterpillars feed on passion vines in the family Passifloraceae, particularly the genus *Passiflora*, that live scattered through the forest, and Gilbert (1975, 1982) has suggested that herbivory by *Heliconius* caterpillars has been a major selective force on the evolution of leaf shape in these vines. The butterflies are very host-specific, and adult females rely strongly on vision to locate suitable vines. Caterpillar attack can significantly reduce the growth and performance of the passion vines. Given these conditions, Gilbert suggests that three elements in the morphology of the leaves of these plants may have evolved in response to selection imposed by *Heliconius*. They are mimicry, divergence in leaf shape between species, and markedly different adult and juvenile foliage.

The case for mimicry rests on the observation that in many habitats, leaves of passion vines closely resemble, in both shape and texture, leaves of abundant, non-host plants. Models include *Philodendron* and *Rubus*. Gilbert suggests that an egg-laying female, searching visually, could easily overlook suitable, mimetic hosts. Second, in many areas where different species of passion vines coexist, they often differ markedly from one another in leaf shape, even though they are closely related and share similar physical conditions. Gilbert sees divergence in leaf shape as a means of reducing attack, because ovipositing *Heliconius* overlook potential hosts with the wrong leaf shape. (Consistent with this hypothesis, caterpillars of some species can feed and develop on the 'wrong' hosts, if introduced onto them experimentally (Smiley 1978, 1985).) Divergence in leaf shape between species can therefore be viewed as an example of selection for 'enemy-free space' (Jeffries & Lawton 1984). Finally, differences between adult and juvenile foliage within one species of vine, Gilbert argues, has also evolved in response to selection from *Heliconius*. Juvenile foliage may mimic other plants on the forest floor, and altering the appearance of the leaves as the plant matures may further reduce the chances of female butterflies recognizing vulnerable juveniles.

Gilbert himself is the first to admit the speculative nature of these arguments. As in the case of Australian mistletoes, one could postulate, for example, that convergence in leaf size and shape is a response to shared physical environmental variables, rather than mimicry. It is also possible that different adult and juvenile foliages are adaptations to different physical environments on the shade of the forest floor and in the canopy. More work is required to distinguish between the herbivore hypothesis and the physiological and biomechanical hypotheses that have so far dominated thinking about the design of leaves.

5. INTERACTIONS BETWEEN OTHER INSECT HERBIVORES AND LEAF SIZE AND SHAPE

A necessary, although not sufficient, condition for insects to influence the design of leaves is that herbivores must respond to, and be affected by, leaf size and shape. Then, if these responses ultimately change herbivore abundances and hence damage-levels experienced by the plant, the way is open for insect herbivores to influence the evolution of leaf size and shape. Aside from the *Heliconius–Passiflora* system discussed above, there is sufficient evidence scattered through the literature to make a case for other insects as selective agents on the size and shape of leaves. We have gathered the arguments together under five headings, namely that leaf size and shape are involved in host-plant recognition and selection by insects; they can modify the efficiency with which herbivores exploit leaves; certain leaf shapes may act as physical barriers to insect herbivory; leaf size and shape also influence the number of species of herbivores attacking plants; and sometimes, at least, they alter herbivore abundances.

An added, but very poorly studied, possibility is that leaf size and shape have similar effects on the parasitoids and predators of herbivores, thereby changing herbivore abundances via the third trophic level. Insect herbivores may also be better camouflaged and concealed from enemies on leaves of certain sizes and shapes. There is no explicit discussion of these problems in the major review of plant–herbivore–enemy interactions undertaken by Price *et al.* (1980); we touch on some possible examples.

(a) *Leaf size and shape influence selection of plants by insects*

Both psilids (Diptera) and butterflies (Lepidoptera) are attracted towards plants bearing leaves of a particular shape (Prokopy & Owens 1983). Well studied examples are provided by ovipositing female butterflies searching for hosts; they include not only the *Heliconius* spp. discussed above, but also *Battus philenor* (Rausher 1978) and two species of *Eurema* (Mackay & Jones 1989). Adults of the fly *Pegomya nigritarsis* (Diptera) can gauge leaf size because they adjust the number of eggs laid according to the sizes of the leaves of their *Rumex* hosts (Godfray 1986), although they may do this using chemical cues correlated with leaf size, rather than vision. Aphids

(Homoptera), too, may be sensitive to leaf size; adult females of *Pemphigus betae* seek out and preferentially colonize large leaves of *Populus angustifolia* (Whitham 1978), with marked effects on aphid fitness (see below). This is not to say that any of these herbivores have influenced the evolution of leaf size or shape in these particular plants; the examples simply illustrate that herbivores are sensitive to gross aspects of leaf morphology.

Evidence that some insects recognize and respond to leaf shape prompted Niemela & Tuomi (1987) to suggest that narrow, irregular hollows on the leaf blades of certain Moraceae mimic caterpillar feeding damage. These irregular 'incisions' occur on some of a plant's leaves, but not others, enhancing resemblance to feeding damage. Niemela & Tuomi speculate that false damage could protect the plant in several ways. Ovipositing females may avoid it because real damage signals the presence of competing larvae, or induced biochemical defences in leaves; false damage may also be avoided because real damage serves to attract parasitoids and predators. Niemela & Tuomi point out that if natural enemies are attracted to false damage, thereby increasing herbivore mortality rates, selection may also operate on leaf shape via the third trophic level. A wide range of plants outside the Moraceae also have leaves with herbivore-like indentations and holes (M. J. Crawley, personal communication; V. K. Brown & J. H. Lawton, personal observations). The mature, but not the juvenile, leaves of *Monstera deliciosa* are a familiar example. The hypothesis that 'pseudo-damage' is mimetic and protective against herbivores (including vertebrates), deserves more attention, along with the related suggestion that variegated colour patterns on certain leaves may serve a similar function, for example by mimicking leaf-mines (Smith 1986; but see Givnish (1990)).

(b) Leaf size and shape influence the performance of individual insects

It would be remarkable if the size and shape of leaves did not influence the lives of insects living on and in them. Movement, mating, oviposition, feeding, hiding and sheltering by both herbivores and their enemies must all be affected by the design of leaves (see, for example, Heinrich (1971)). Surprisingly, the problem has received little systematic attention. Leaf-surface properties (roughness, waxiness, pubescence, etc.) influence the ability of insects to hold on, move about and feed (see Strong *et al.* 1984), but lie outside the scope of this paper. Nevertheless, there are data showing that more macroscopic properties of leaves also impinge on many aspects of the lives of insects.

Female *Lithocolletis quercus* (Lepidoptera) preferentially oviposit in large leaves, although discrimination is not very acute (Auerbach & Simberloff 1989), and how they judge leaf size is unclear. In contrast, female *Perga affinis* (Hymenoptera: Symphyta) cannot oviposit on wide-leaved hosts, because they need to grip the leaf edges when laying (Carne 1962). Feeding efficiency in adult Christmas beetles, *Anoplognathus chloropyrus* (Coleoptera), is similarly re-

duced on narrow *Eucalyptus* leaves because they cannot manoeuvre effectively (Carne *et al.* 1974). The juvenile leaves of blue gums lack petioles, a characteristic (together with the fact that they are softer than adult foliage) that makes them particularly susceptible to attack by caterpillars of the autumn gum moth, *Mnesampela privata* (Lepidoptera); the caterpillars can more easily withdraw into protective daytime shelters on leaves without petioles (Elliot & Bashford 1978). Conversely, lepidopteran caterpillars may find the small and often sparse leaflets on the pinnate leaves of African *Acacias* difficult to manoeuvre and eat; leaflets are also very loosely attached and often fall off as caterpillars try to grasp them (M. Coe, personal communication). Exploitation of some pinnate leaves by caterpillars may also be made more difficult, and the risk of herbivore damage to the petiole reduced, because basal leaflets are less palatable than lateral or terminal leaflets (Gall 1987).

Finally, at least three measures of reproductive success in *Pemphigus* aphids are correlated with the size of the mature leaf upon which they are feeding (Whitham 1978). In comparisons between large (more than 15 cm long) and small (up to 5 cm long) leaves on the same branch, aphids survived better (0 versus 80 % mortality), were 70 % larger, and gave birth to 220 % more progeny on large leaves.

We are surprised that more data do not exist on the behaviour and performance of individual insects exploiting different sizes and shapes of leaves. Three sorts of data would be instructive.

1. Observations on polyphagous species moving about a range of their normal host plants that differ in leaf size and shape. Heinrich (1971) provides a pioneering example.

2. Comparisons of the performance of herbivores (and natural enemies) on single species of plants with variable shaped leaves, for example the dissected and pinnatifid cultivars of some *Betula*, *Fagus* and *Acer* spp. compared with normal phenotypes.

3. Information on the performance and survival of leaf miners and gall formers on leaves of different sizes and shapes.

(c) Leaves as physical barriers to insect herbivores

Leaves may present physical barriers to insect herbivores in other ways than being tough, or by having hairs, trichomes, and marginal teeth or spines. *Blackstonia perfoliata*, a European member of the Gentianaceae, has connate stem leaves. By surrounding the stem, these leaves present a formidable barrier to any small insect trying to climb the plant, reminiscent of the discs on the ropes of ships to keep rats ashore. The juvenile foliage of some eucalypts is very similar. Do leaves that completely surround the stem function as 'rat stops' for herbivores seeking terminal buds?

(d) Effects of leaf size and shape on herbivore species richness

British umbellifers (Apiaceae) have leaf shapes ranging from finely divided, filiform leaves to the broad, and simply pinnate. Controlling for size of host-plant geographic range (which has a large influence on herbivore species richness), umbellifers with finely divided leaves support significantly fewer species of agromyzid leaf-miners (Diptera) (Lawton & Price 1979), and significantly fewer species of insect herbivores in general (Jones & Lawton 1991) than umbellifers with large, undivided leaves. Although the mechanism is not well understood, it presumably rests on the difficulties herbivores experience in exploiting more finely divided leaves. There must, for example, be a critical width of tissue and a critical size for individual leaflets, below which leaf-mining is impossible, and external feeding becomes inefficient. Trees yield similar results, with fewer insect species exploiting species with smaller leaves (Moran & Southwood 1982; Kennedy & Southwood 1984).

Subtle effects on different guilds are revealed by comparisons of herbivore assemblages on two vetches (Leguminosae) both with pinnate leaves, *Vicia hirsuta* (with very narrow leaflets) and *V. sativa* (with broader leaflets); *hirsuta* hosts fewer species of external chewing herbivores than *sativa*, but has more sucking (sap-feeding) species (Brown *et al.* 1987).

(e) Effects on population dynamics and abundance of insects

Rates of population growth of pea aphids, *Acyrthosiphon pisum*, are significantly lower on cultivars of peas lacking leaves than on normal plants (Kareiva & Sahakian 1990). Because the leafless cultivars have tendrils, they are analogues of plants with exceptionally finely divided, filiform leaves. Aphids perform better on normal plants, because ladybird predators (Coccinellidae) fell off the normal variety about twice as often as they did from the leafless variety, reducing predation rates on normal plants; beetles can cling to stems and tendrils much better than to smooth, flat leaves.

Absolute abundances of herbivores are difficult to compare between plants with markedly different sizes and shapes of leaves, unless scaled for leaf area. In insecticide-knockdown samples collected in 1 m² sheets (which may roughly control for leaf area), narrow leaved trees supported only about one quarter the number of individual insects fogged from species with broader leaves (Moran & Southwood 1982). Brown *et al.* (1990) and Sterling *et al.* (1991) found that sheep grazing modified the sizes of leaves of *Ranunculus repens* and *Medicago lupulina*. Two species of leaf-mining flies (Diptera) were more abundant in grazing treatments, where host plants had larger leaves, even though the absolute abundance of their food plants was much reduced. Similarly, P. Niemela (personal communication) found more eriophyid (Acari) galls per unit area on large, palmate leaves than on smaller ovate leaves.

(f) Insect herbivores: an overview

Drawing these arguments together, there is scattered but growing evidence showing that insect herbivores (and their enemies) respond to and are affected by leaf size and shape, generating differences in herbivore species richness, or herbivore abundances, on different species or genotypes of plants with different sizes and shapes of leaves. In general, small or highly divided leaves appear more difficult for insect herbivores to exploit than large, entire leaves. However, there are as yet no direct demonstrations that herbivore responses to leaf size and shape translate into effects on plant fitness, or plant performance, so that the case for insect herbivores as significant players in the evolution of leaf size and shape is circumstantial.

6. CONCLUDING REMARKS

There are at least six ways in which herbivores may have influenced the evolution of leaf size and shape, favouring leaf morphologies that differ from those dictated by physiological and biomechanical constraints acting on plants. They are mimicry, not only of leaves of other species of plants but also grazed leaves and inanimate objects; crypsis; physical barriers to being eaten; interspecific differences in leaf morphology to reduce recognition by herbivores; very small or highly divided and dissected leaves that reduce feeding efficiency; and different adult and juvenile foliage.

We do not know how common most of these responses are. Nor do we know the extent to which carbon gain by individual plants is compromised below physiological optima by changes in leaf size and shape that serve to reduce herbivory. Grazing jeopardizes the plant in several ways. As well as the initial loss of biomass, the future photosynthetic contribution of the grazed tissue is also lost (Chabot & Hicks 1982), and may be exacerbated by premature abscission of damaged leaves. Wounds are also vulnerable to invasion by pathogens. Selection may therefore favour leaf sizes and shapes that differ from physiological and biomechanical optima, providing such leaves bring compensating gains to the plant via reduced risks of attack by herbivores (Givnish 1979). It is currently unclear whether herbivory is more likely to influence the evolution of leaf size and shape in some habitats than others; or whether certain types of plants, for instance particularly palatable species high in nitrogen, are more or less likely to respond to herbivore attack by changes in leaf macro-morphology (see Givnish (1990) for a brief discussion). The problem is further complicated by the need to consider the costs and benefits of alternative means of defence open to the plant, not least allelochemicals. There are, however, no reasons to believe that changes in leaf size and shape are any more difficult to evolve than other aspects of plant morphology or biochemistry (Gottlieb 1984).

Insights into the evolutionary impact of herbivores can be gained by studying plant morphology in the absence of key players, mammals in New Zealand for instance. An alternative would be to examine leaf size and shape in early angiosperm floras, before the evolution of herbivorous mammals (Collinson & Hooker, this symposium), or significant numbers of extant insect herbivores (for examples, see Strong *et al.* (1984); Chaloner *et al.* (this symposium)). To our untutored eyes, the fossil leaves of many early angiosperms have a very simple, rather uniform morphology (see, for example, Hickey & Wolfe (1975); Hughes (1976)), although pinnate and palmate forms appear early in some Cretaceous floras (Hallam 1977), and it may be difficult to separate the role of herbivores from the effects of palaeo-climate. Comparisons between floras, both extant and extinct, would be easier if the range and variety of leaf sizes and shapes could be quantified objectively. It would be particularly valuable to have data on leaf sizes and shapes, gathered in a standard way, for representative samples of plant communities in different biomes and biogeographic regions.

We have the impression that in some habitats plants have leaves of a rather uniform size and shape, whereas in others leaves of different species span a much greater range of morphologies. Why? Is it that certain habitats are physiologically demanding and strongly constrain leaf design, or is it because herbivore pressures are very different in different habitats? The answer is probably both. Speculating, we predict that habitats where visually searching herbivores are common will display a greater range and variety of leaf sizes and shapes than habitats where such herbivores are rare or absent. As well as selecting for mimicry in plants, visually searching herbivores may also impose strong selection for leaves of different species of plants to be different shapes, to reduce risks of being recognized as food. There may often be nothing special about particular shapes of leaves; only that they should be different from leaves of other species in the same habitat.

There are a number of ways in which an assessment of the role of herbivores in the evolution of leaf size and shape could be made more rigorous. One is to exploit genetic and phenotypic differences in leaf morphology between individual plants, or between adult and juvenile foliage or between normal and regrowth leaves, in experiments that measure the performances of both plants and herbivores. Without such experiments, an assessment of the role of herbivores in the evolution of leaves will always be difficult. Our view is that herbivores have played a significant role in determining the sizes and shapes of leaves, and that without herbivores, leaf morphology would be simpler and less interesting.

We are extremely grateful to Bill Chaloner, Malcolme Coe, Mick Crawley, Penny Edwards, John Harper, Matt McGlone, Pekka Niemela, Dave Parkhurst and John Richards for ideas and valuable discussions, and to Matt McGlone and Sue Milton for permission to quote from manuscripts prior to publication.

REFERENCES

Atkinson, I. E. A. & Greenwood, R. M. 1989 Relationships between moas and plants. *N.Z. Jl Ecol.* **12** (suppl.), 67–96.

Atsatt, P. R. 1981 Lycaenid butterflies and ants: selection for enemy-free space. *Am. Nat.* **118**, 638–654.

Atsatt, P. R. 1983 Mistletoe leaf shape: a host morphogen hypothesis. In *The biology of mistletoes* (ed. M. Calder & P. Bernhardt), pp. 259–275. Sydney: Academic Press.

Auerbach, M. & Simberloff, D. 1989 Oviposition site prefer-
ence and larval mortality in a leaf-mining moth. *Ecol. Ent.* **14**, 131–141.

Barlow, B. A. & Wiens, D. 1977 Host-parasite resemblance in Australian mistletoes: the case for cryptic mimicry. *Evolution* **31**, 69–84.

Barrett, S. C. H. 1983 Crop mimicry in weeds. *Econ. Bot.* **37**, 255–282.

Böcher, T. W. & Lewis, M. C. 1962 Experimental and cytological studies on plant species VII. *Geranium sanguineum. Biol. Skr. Dan. Vid. Selsk.* **11** (5), 1–25.

Brown, V. K., Gange, A. C., Evans, I. M. & Storr, A. L. 1987 The effect of insect herbivory on the growth and reproduction of two annual *Vicia* species at different stages in plant succession. *J. Ecol.* **75**, 1173–1189.

Brown, V. K., Gibson, C. W. D. & Sterling, P. H. 1990 The mechanisms controlling insect diversity in calcareous grasslands. In *Calcareous grasslands – ecology and management* (ed. S. H. Hillier, D. W. H. Walton & D. A. Wells), pp. 79–87. Huntingdon: Bluntisham.

Carne, P. B. 1962 The characteristics and behaviour of the saw-fly *Perga affinis* (Hymenoptera). *Aust. J. Zool.* **10**, 1–34.

Carne, P. B., Greaves, R. T. G. & McInnes, R. S. 1974 Insect damage to plantation-growth eucalypts in north coastal New South Wales, with particular reference to Christmas beetles (Coleoptera: Scarabaeidae). *J. Aust. Entom. Soc.* **13**, 189–206.

Chabot, B. F. & Hicks, D. J. 1982 The ecology of leaf life spans. *A. Rev. Ecol. Syst.* **13**, 229–259.

Ehrlich, P. R. & Raven, P. H. 1967 Butterflies and plants. *Scient. Am.* **216** (6), 104–113.

Ehleringer, J. R., Ullmann, I., Lange, O. L., Farquhar, G. D., Cowan, I. R. & Schulze, E.-D. 1986 Mistletoes: a hypothesis concerning morphological and chemical avoidance of herbivory. *Oecologia* **70**, 234–237.

Elliot, H. T. & Bashford, R. 1978 The life history of *Mnesampela privata* (Guan.) (Lepidoptera: Geometridae) a defoliator of young eucalypts. *J. Aust. Entom. Soc.* **17**, 201–204.

Everard, B. & Morley, B. D. 1970 *Wild flowers of the world.* London: Ebury Press & Michael Joseph.

Gall, L. F. 1987 Leaflet position influences caterpillar feeding and development. *Oikos* **49**, 172–176.

Gilbert, L. E. 1975 Ecological consequences of a coevolved mutualism between butterflies and plants. In *Coevolution of animals and plants* (ed. L. E. Gilbert & P. R. Raven), pp. 210–240. Austin: University of Texas Press.

Gilbert, L. E. 1982 The coevolution of a butterfly and vine. *Scient. Am.* **247** (August 1882), 102–107.

Givnish, T. J. 1979 On the adaptive significance of leaf form. In *Topics in plant population biology* (ed. O. T. Solbrig, S. Jain, G. B. Johnson & P. H. Raven), pp. 375–407. London: Macmillan.

Givnish, T. J. 1986 (ed.) *On the economy of plant form and function.* Cambridge University Press.

Givnish, T. J. 1990 Leaf mottling: relation to growth form and leaf phenology and possible role as camoflage. *Funct. Ecol.* **4**, 463–474.

Godfray, H. C. J. 1986 Clutch size in a leaf-mining fly (*Pegomya nigritarsis*: Anthomyidae). *Ecol. Ent.* **11**, 75–81.

Gottlieb, L. D. 1984 Genetics and morphological evolution of plants. *Am. Nat.* **123**, 681–709.

Hallam, A. 1977 (ed.) *Patterns of evolution as illustrated by the fossil record.* Amsterdam: Elsevier Scientific.

Heinrich, B. 1971 The effect of leaf geometry on the feeding behaviour of the caterpillar of *Manduca sexta* (Sphingidae). *Anim. Behav.* **19**, 119–124.

Hickey, L. J. & Wolfe, J. A. 1975 The bases of angiosperm phylogeny: vegetative morphology. *Ann. Missouri Bot. Gard.* **62**, 538–589.

Horn, H. S. 1971 *The adaptive geometry of trees.* Princeton University Press.

Hughes, N. F. 1976 *Palaeobiology of angiosperm origins.* Cambridge University Press.

Jeffries, M. J. & Lawton, J. H. 1984 Enemy free space and the structure of ecological communities. *Biol. J. Linn. Soc.* **23**, 269–286.

Jones, C. G. & Lawton, J. H. 1991 Plant chemistry and insect species richness of British umbellifers. *J. Anim. Ecol.* **60**. (In the press.)

Kareiva, P. & Sahakian, R. 1990 Tritrophic effects of a simple architectural mutation in pea plants. *Nature, Lond.* **345**, 433–434.

Kennedy, C. E. J. & Southwood, T. R. E. 1984 The number of species of insects associated with British trees: a re-analysis. *J. Anim. Ecol.* **53**, 455–478.

Lawton, J. H. & Price, P. W. 1979 Species richness of parasites on hosts: agromyzid flies on British Umbelliferae. *J. Anim. Ecol.* **48**, 619–637.

Mackay, D. A. & Jones, R. E. 1989 Leaf shape and host-finding behaviour of two ovipositing monophagous butterfly species. *Ecol. Ent.* **14**, 423–431.

McGlone, M. S. 1991 A critical examination of the moa-browsing hypothesis for the origin of the divaricating plant syndrome. (Submitted.)

McGlone, M. S. & Webb, C. J. 1981 Selective forces influencing the evolution of divaricating plants. *N.Z. Jl Ecol.* **4**, 20–28.

Milton, S. J. 1991 Plant spinescence in arid southern Africa: does moisture mediate selection by mammals? *Oecologia.* (In the press.)

Moran, V. C. & Southwood, T. R. E. 1982 The guild composition of arthropod communities on trees. *J. Anim. Ecol.* **51**, 289–306.

Niemela, P. & Tuomi, J. 1987 Does the leaf morphology of some plants mimic caterpillar damage? *Oikos* **50**, 256–257.

Parkhurst, D. F. & Loucks, O. L. 1972 Optimal leaf size in relation to environment. *J. Ecol.* **60**, 505–537.

Price, P. W., Bouton, C. E., Gross, P., McPheron, B. A., Thompson, J. N. & Weis, A. E. 1980 Interactions among three trophic levels: influence of plants on interactions between insect herbivores and natural enemies. *A. Rev. Ecol. Syst.* **11**, 41–65.

Prokopy, R. J. & Owens, E. D. 1983 Visual detection of plants by herbivorous insects. *A. Rev. Ent.* **28**, 337–364.

Rausher, M. D. 1978 Search image for leaf shape in a butterfly. *Science, Wash.* **200**, 1071–1073.

Smiley, J. 1978 Plant chemistry and the evolution of host specificity: new evidence from *Heliconius* and *Passiflora*. *Science, Wash.* **201**, 745–747.

Smiley, J. T. 1985 Are chemical barriers necessary for evolution for butterfly-plant associations? *Oecologia* **65**, 580–583.

Smith, A. P. 1986 Ecology of leaf color polymorphism in a tropical forest species: habitat segregation and herbivory. *Oecologia* **69**, 283–287.

Southwood, T. R. E. 1987 Plant variety and its interaction with herbivorous insects. In *Insects – plants* (ed. V. Labeyrie, G. Fabres & D. Lachaise), pp. 61–69. Dordrecht: Junk.

Sterling, P. H., Gibson, C. W. D. & Brown, V. K. 1991 Leaf miner assemblies: effects of plant succession and management. (Submitted.)

Strong, D. R., Lawton, J. H. & Southwood, T. R. E. 1984 *Insects on plants. Community patterns and mechanisms.* Oxford: Blackwell Scientific Publications.

Whitham, T. G. 1978 Habitat selection by *Pemphigus* aphids in response to resource limitation and competition. *Ecology* **59**, 1164–1176.

Discussion

P. J. Grubb (*Botany School, University of Cambridge, U.K.*). Two points. First, Givnish was not only one of the first to canvas the idea that mottling of the leaves of some herbs in cool temperate deciduous forests is a form of camouflage against grazing animals (in the dappled light of the forest floor), but also one of the first to emphasize the contrast in such forests between the striking convergence in leaf form among the many herbaceous and woody species that are summer-green and the equally remarkable variety in leaf-form that exists in the species of 'spring ephemeral' whose leaves die down soon after the trees leaf out. If a compendium of all the world's plant species is to be drawn up, including codified records of physiognomy and habitat as suggested by Professor May, it will clearly be important to include records of the phenology (seasonal incidence) of leafing.

Secondly, the remarkable plants the authors have described from New Zealand, with various features apparently selected by grazing pressure from extinct moas, should be seen in perspective. The great majority of native plants in New Zealand have forms that would be expected by an ecologist experienced in the forests and alpine zones of the tropics, subtropics and warm temperate regions elsewhere. The plants 'moulded' by moas are almost all confined to specialized sites: for example, the edges of forest against swamps, etc. (the divaricating shrubs) where the plants were constantly accessible to moas, and tree-fall gaps (narrow-leaved *Pseudopanax*) where they were similarly accessible, and also scarce on the scale of the whole landscape.

On the evolution of plant secondary chemical diversity

CLIVE G. JONES[1] AND RICHARD D. FIRN[2]

[1] *Institute of Ecosystem Studies, Millbrook, New York 12545, U.S.A.*
[2] *Department of Biology, University of York, Heslington, York YO1 5DD, U.K.*

SUMMARY

A common-sense evolutionary scenario predicts that well-defended plants should have a moderate diversity of secondary compounds with high biological activity. We contend that plants actually contain a very high diversity of mostly inactive secondary compounds. These patterns result because compounds arising via mutation have an inherently low probability of possessing any biological activity. Only those plants that make a lot of compounds will be well defended because only high diversity confers a reasonable probability of producing active compounds. Inactive compounds are retained, not eliminated, because they increase the probability of producing new active compounds. Plants should therefore have predictable metabolic traits maximizing secondary chemical diversity while minimizing cost. Our hypothesis has important implications to the study of the evolution of plant defence.

1. EVOLUTION OF WELL-DEFINED PLANTS: A COMMON-SENSE SCENARIO

Commonly accepted facts and notions about plant secondary metabolites can be used to construct a common-sense scenario for the evolution of well-defended plants. Compounds arising via enzyme mutations adversely affect herbivores and plant pathogens (Rosenthal & Janzen 1979; Horsfall & Cowling 1980). A diversity of active compounds confers greatest resistance because mixtures of compounds enhance activity and because plants interact with many organisms (Jones 1983; Berenbaum 1985). Because production of compounds incurs energy, material and opportunity costs (McKey 1979; Rhoades 1979), diversification is constrained once a plant is well defended. At the same time, consumer adaptation (i.e. avoidance, detoxification and utilization of compounds; Chapman & Blaney 1979; Brattsten 1979; Bernays & Woodhead 1982) and new colonists will tend to decrease activity of compounds and overall defensive efficacy, with highly active compounds selecting for rapid consumer adaptation (Pimentel & Belotti 1976). Plants that eliminate inactive compounds that are not essential precursors of active compounds will incur reduced costs without any decreased benefits. Further mutations produce new, active compounds, replacing those that have been eliminated. If natural selection favours plants that are well defended but minimize cost, well-defended plants should contain a moderate diversity of highly active compounds, and few if any inactive compounds other than essential precursors. Is this what we find?

2. OBSERVED PATTERNS OF ACTIVITY AND DIVERSITY

(a) *Are most secondary compounds highly active?*

Some highly active compounds are known to occur in plants (e.g. azadirachtin; pyrethroids, nicotine; Jacobson & Crosby 1971). However, most active compounds are rarely 100% lethal or inhibitory at naturally occurring concentrations. Although it is likely that most plants contain a few active compounds, the number of active compounds is a small percentage of all the compounds that have been isolated. Programmes screening plant extracts or pure natural products have a very low frequency of discovery of highly active compounds, irrespective of whether the screening was against consumers or other targets (i.e. cancer–anticancer, medicinal, plant growth regulator). It is a 'basic dogma of random screening that only a very small percent of samples should be active' (Suffness & Douros 1979, p. 81; Goodman & Gilman 1965). For example, in the National Cancer Institute anticancer screening programme, only 4.3% of plant species had any activity, and pure compounds with high activity resulted from 0.07% of species (Suffness & Douros 1982). Rohm and Hass estimate that 1–2% of plants have activity (C. Swithenbank, personal communication), whereas Rothhamsted programmes find about 3% of plants to be active against insects (J. A. Pickett, personal communication). Of 7000 isolates of *Bacillus thuringiensis* from soil that were screened against grass grub, only 7% had any detectable activity and only 0.07% had what could be termed high activity (P. Wigley & C. Chilcott, personal communication).

The synthetic organic chemist is no better at finding activity by screening. Commercial programs are constrained by requirements other than activity *per se*; nevertheless, Dupont estimates that about 0.005% of compounds screened for insecticidal, developmental or antifeedant activities result in a commercial product (J. Frazier, personal communication). Ciba-Geigy observe that random or directed screening with synthetic and natural products have about the same very low success rate (H. Fischer, personal communication; Geissbühler *et al.* 1982).

(b) *Why are so many compounds inactive and why are they retained?*

Is the prevalence of inactivity solely a consequence of rapid consumer adaptation? We think not. Screening programmes with synthetic or natural compounds never encountered by particular consumers, and screening of natural products for biological activities not relevant to consumers (e.g. plant-growth regulators) reveal the same pattern of a low frequency of occurrence of biological activity.

If most compounds are relatively inactive, why are they not eliminated? The common-sense scenario predicts elimination of inactive, inessential (non-precursor) compounds because they incur cost with no benefit. Does retention of inactive compounds mean that the cost of defence is non-existent or unimportant? This seems unlikely, given the many cogent arguments and circumstantial lines of evidence to support a notion of cost, even if we do not know how great the cost is (e.g. trade-offs between inherent growth rate and resource availability, defence and growth (Bryant *et al.* 1983; Coley *et al.* 1985); inducible defences (Tallamy & Raupp 1991)).

(c) *Why are there so many secondary compounds?*

Most plants contain a very high diversity of secondary metabolites. For example, both carrot and bracken fern have 12 biosynthetically distinct pathways, and bracken has at least 27 different structures within the class sesquiterpene indan-1-ones (Jones 1983; Jones & Lawton 1991). It is reasonable to assume that the observed diversity is necessary to be well defended (Haldane 1949). Nevertheless, it appears that plants contain many more compounds than indicated by the common-sense scenario. If true, why do plants produce so many compounds rather than a few? How has this diversity arisen? How is it maintained? How does this diversity relate to the observation that most compounds are inactive?

3. THE INHERENT PROBABILITY OF BIOLOGICAL ACTIVITY

We believe that the discrepancies between expected and observed activity and diversity in the common-sense scenario are due to a hidden assumption that is incorrect. Compounds arising via mutation are assumed to have an inherently high probability of possessing biological activity against organisms with which the plant interacts. Instead, we postulate that the inherent probability is low. This is a fundamental toxicological property of the interaction between chemicals and organisms. The low success rate reported from screening programmes shows that toxicologists are acutely aware of a phenomenon that plants have had to deal with for millions of years. Although it is

interesting to speculate why low activity is a fundamental toxicological property, it is not necessary to know the reasons in order to examine the consequences to plant defence.

4. AN EVOLUTIONARY SCENARIO BASED ON A LOW PROBABILITY OF ACTIVITY

If secondary metabolites have an inherently low probability of possessing biological activity, mutations that add active compounds and increase defence will be rare. Plants producing only inactive compounds incur costs with no defensive benefit and should be selected against. Some plants will have a few active compounds and be well defended. Consumer adaptation and new colonists will then decrease defensive efficacy. As the probability of producing new active compounds is low, how can plants maintain adequate defence in the face of consumer adaptation and colonization?

Consider what would happen if a well-defended plant with a few active compounds retains some inactive compounds. This could occur if the rate of enzyme mutations adding compounds temporarily exceeds the rate of deletion of compounds, and if the cost of retaining these compounds is not excessive (i.e. cost–benefit remains within selectively advantageous limits). The total number of compounds added or deleted over time will be a function of the mutation rate per biosynthetic step (compound) multiplied by the total number of biosynthetic steps (compounds). Thus plants with a high diversity of compounds, irrespective of their activity, have a greater probability of adding or deleting compounds over time than plants with a low diversity of compounds. Because the probability of any compound having activity at any time is low in both cases, it is clear that plants with a high absolute diversity have a greater probability of producing one or more active compounds at any time than plants with a low diversity.

Plants with a high diversity of compounds will 'screen' a greater number of compounds at any time and a greater variety of compounds over time than plants with a low diversity. We argue that this is the only effective 'strategy' if the 'screening success' is inherently low. It is also a 'strategy' that produces the observed patterns of well-defended plants containing a high diversity of mostly inactive compounds and a few active compounds.

Maintenance of high diversity therefore increases the probability of production of new active compounds in the face of consumer adaptation. A high diversity can also increase the probability that selection for consumer adaptation to one active compound may then lead to susceptibility to another, previously inactive compound (see, for example, Gould 1988). Furthermore, because plants are continually subject to attempted colonizations by new consumers (Jones 1983; Strong *et al.* 1984), plants with a high diversity of compounds also have a higher probability of possessing a compound active against potential colonists.

5. METABOLIC TRAITS MAXIMIZING DIVERSITY AND MINIMIZING COST

Maintenance of a high diversity of secondary chemicals is therefore advantageous, but it probably incurs costs. We would therefore expect natural selection to favour plants that possess metabolic traits that maximize diversity and minimize cost. What would such traits look like and do plants possess them?

(a) Branched pathways

Branched pathways of secondary metabolism should be relatively common and simple linear pathways should be rare. Continual bifurcation (i.e. mutation at a biosynthetic step results in production of a new product and retention of the existing product) results in exponential doubling in the number of compounds over time at a constant mutation rate per biosynthetic step, compared with additive increases in diversity with linear pathways (i.e. mutation at a biosynthetic step results in production of a new product and deletion of the existing product). For example, after only four mutations a bifurcating pathway will have 1.8 times as many compounds as a linear pathway. Deletion mutations are likely to have less of an effect on diversity in complex versus simple pathways. For example, in a linear pathway with six biosynthetic steps (i.e. seven compounds), a single random deletion mutation has an equal probability of removing either one, two, three, four, five or six compounds. However, in a bifurcating pathway with the same number of steps and compounds, a single mutation has four chances of removing only one compound and two chances of removing three compounds. Similarly, an active compound may have a lower probability of being eliminated by a single mutation in a bifurcating than in a linear pathway. Thus in the above example, an active compound at the end of the linear pathway will be removed by a deletion at any of the six steps; whereas only two of six possible deletions will eliminate the active compound if it is at one of the four terminal points of a bifurcating pathway.

Highly branched pathways are one of the most striking characteristics of secondary metabolism (Mann 1987). For example, pathways for biosynthesis of sesquiterpenes, alkaloids, anthraquinones and shikimate derivatives are all highly branched (Bu'lock 1965; Iwasaki & Nozoe 1974).

There are two further consequences of branched pathways. First, as the diversity of a pathway increases, a greater proportion of the total diversity is found in any one pathway. Thus, diversity of compounds within a class is always likely to exceed that between classes, as is observed (e.g. bracken; Jones 1983). Second, a new mutation is far more likely to add compounds to an existing pathway than start a new pathway from a primary metabolite, simply because the number of biosynthetic steps located in existing pathways is much greater. Consequently, there will be a high degree of canalization in pathway diversification, leading to the high degree of gross chemotaxonomic affinity that we observe in plants, despite the fact that there are clear differences in specific compound structure between related plants (see Harborne 1977).

(b) Matrices and grids

Metabolic matrices and grids should be common. Here a few enzymes convert a diversity of structurally similar precursors to many end products. Such matrices and grids confer the same advantages as bifurcating pathways, but can also result in substantial cost savings. For example, in a four-step matrix with four precursors, 20 compounds are produced with only four enzymes. Matrices and grids are common in secondary metabolism. For example, biosynthesis of polyketides, carotenoids, xanthophylls, flavanoids, coumestrols, and alkaloids; cinnamic acid conversions and formation of coumarins and cinnamoyl quinic acids all involve grids and matrices (Bu'lock 1965; Hahlbrock & Grisebach 1975).

(c) Combining pathways

Combination of different pathways to produce new products is another means of increasing diversity while minimizing cost. Given an existing cost commitment to produce two pathways, biosynthesis involving products from each pathway will further enhance diversity at relatively little extra cost. Compounds of mixed biosynthetic origin are common. For example, mycophenolic acids, cannabinoids, isoprenoid quinones, furanocoumarins, furanoquinolines, alizarin anthraquinones, and certain flavanoids and alkaloids are all formed this way (Mann 1987).

(d) Pathway regulation

Most compounds in secondary metabolic pathways are produced in trace amounts, with one or a few components being quantitatively dominant (e.g. bracken sesquiterpenes, Jones & Firn 1979; monoterpenes, Gershenzon 1991). Major and minor components can occur in concentrations that differ by three or four orders of magnitude. Minor components have the potential to be just as active as a major components because the doses at which different compounds affect the same organism, and the dose at which the same compound affects different organisms can both vary by orders of magnitude. Thus a plant may obtain the same degree of defence from less costly minor components as more costly major components. Furthermore, mutations in the regulation of pathways, as a result of alterations in promoter or regulator genes and enzyme–substrate affinities, can potentially result in a new quantitative profile. Major components could become minor components and vice versa. This could result in marked shifts in biological activity with no real change in cost. The fact that closely related species or genotypes can possess the same compounds but have substantially different quantitative profiles (e.g. monoterpenes in conifers; Cates & Redak 1988) indicates that such shifts are relatively common.

Similarly, pathway regulation may be altered so that virtually all of the compounds are expressed in trace amounts. This maintains diversity and the potential for activity, while reducing the material and energy costs. The phytoecdysteroids of bracken fern that we previously termed 'redundant defences' (Jones & Firn 1978) may conform to this pattern. A logical extension of this idea is inducible pathways. Here compounds are only produced following a particular signal associated with consumer attack, and operate over relatively short time frames. Such inducible pathways may substantially reduce material and energy costs without jeopardizing defensive capacity. Inducible pathways are widespread in plants (Tallamy & Raupp 1991).

(e) Hidden pathways

It is possible that mutations to regulator or promoter genes, or any type of mutation that deletes the first step in a pathway, could result in an entire pathway no longer being expressed, with substantial savings in energy and material costs. However, the genetic or biosynthetic machinery still remains, with the capacity to re-express the entire pathway after subsequent mutations. Plants may therefore contain hidden pathways. Carrot, a well-studied plant, produces minute traces of methyl sulphides, compounds characteristic of the totally unrelated onion family, and not reported for any other umbellifer (Jones & Lawton 1991). This suggests that carrot has the capacity to produce onion compounds, given a suitable regulatory mutation. We suspect that this type of phenomenon will turn out to be widespread in plants. If true, it would indicate that phenotypic characteristics of secondary metabolism may grossly underestimate the genotypic potential for chemical diversity. The observed chemotaxonomic affinities of plants may be just as much a consequence of failure to express pathways common to many unrelated taxa, as it is a consequence of possessing pathways only found in related taxa.

(f) Reaction mechanisms

Several other metabolic traits related to conversion reactions could promote diversification or decrease cost. First, we would expect that many of the enzymes of secondary metabolism would show relatively low substrate specificity. Such 'sloppy enzymes' could convert a range of structurally similar substrates into slightly different products; or do the same functional transformations on structurally dissimilar substrates, such as hydroxylation or methoxylation. Here the cost of many different transformations is reduced because only one enzyme is involved. Second, we would expect many isozymes. Mutations in genes coding for a particular enzyme could result in an isozyme with reduced specificity that then converts substrates that were not previously utilized, or produces different products from the same substrate. Isozymes of enzymes that do more than one transformation might result in products where only one of these transformations had occurred. We would expect 'sloppy enzymes' and isozymes to be important in branched pathways (see §5a). There is some evidence for sloppy substrate specificity and isozymes, particularly in metabolic matrices and grids (Bu'lock 1965); biosynthesis of

phenolics (Neish 1964; Hahlbrock & Grisebach 1975) and reticuline and other alkaloids; oxidation of carotenoids, polyketide transformations, flavanoid hydroxylation and polymeric isoprenoid formation (Bu'lock 1965).

Third, we would expect polymeric assembly of secondary metabolites. Here, multiple classes of compounds are produced by sequentially combining different numbers of base units, with substantial savings in enzyme costs. Terpene biosynthesis is the classic example of polymeric assembly (Bu'lock 1965).

Lastly, we would expect that non-enzymic, chemical reactions would be important. Here many variants of a structure can be produced without any enzyme costs. Changes in cellular or vacuolar pH and ionic strength, for example, may facilitate such conversions as radical coupling and shiff base formation (Luckner 1972). Monoterpene and phenolic biosynthesis are good examples (Bu'lock 1965; Templeton 1969; Herbert 1981).

6. TESTING THE THEORY

We have presented some evidence in support of our ideas about the relations between plant defence, biological activity, chemical diversity and cost of defence. However, these ideas clearly require further evaluation and testing.

(a) General patterns of diversity and activity

The general patterns that are predicted by our theory are: (i) Most secondary plant compounds will have relatively low activity or no activity against a given organism; highly active compounds will be rare. (ii) Active compounds will have a relatively narrow spectrum of specificity because high activity requires receptor site specificity. (iii) Thus, highly active, broad-spectrum compounds will be very rare. (iv) Most plants will contain a few compounds active against organisms with which the plant interacts and organisms with which the plant has never interacted. (v) Well-defended plants will have both a higher absolute chemical diversity and a greater proportion of active to inactive compounds than poorly defended plants.

There are several problems with testing these general predictions. For example, it is hard to define what is meant by a well- versus a poorly defended plant. If plant defence is a zero-sum game, i.e. only more or less well-defended plants survive, there may be no poorly defended plants available for comparison. Furthermore, it is hard to estimate how many active compounds are necessary for good defence, or how many compounds are needed before an active compound is produced; i.e. there is no clear null model. Nevertheless, there are three approaches that could facilitate testing these general predictions. First, a more thorough analysis of existing data on activity and diversity is needed, to more clearly define patterns.

Second, only a small fraction of all plants and a small fraction of the compounds they contain have been tested against a small fraction of the consumers with which a plant interacts or could interact. Thus, a few comprehensive studies are needed that more fully characterize the chemical diversity of selected plants. Most of the compounds they contain should be screened at naturally occurring concentrations, singly and in admixture, against most of the organisms with which a given plant interacts (i.e. both adapted consumers and potential colonists; for example, see Jones & Firn (1979)).

Third, mathematical or simulation approaches may be of value in generating null models. Such techniques could examine diversity–activity relations as a function of variable values for inherent probability of activity, mutation rates adding and deleting compounds, rates of consumer adaptation and costs of defence.

(b) Metabolic traits

Similar approaches could be taken for examining relations between diversity and activity at the metabolic level. For example, comparison of known pathway structures between primary and secondary metabolism, and between highly diversified and less diversified classes of secondary metabolites might reveal clear differences in the degree of bifurcation or matrix and grid formation, and frequency of pathway recombination. Modelling may facilitate comparisons between actual pathway structures and predicted structure based on maximizing diversity, minimizing the probability of loss of active compounds, and minimizing the cost of production.

There is a major advantage to examining predicted metabolic traits, as opposed to general patterns. Modern techniques of molecular biology could be used to investigate the regulation and expression of secondary metabolism. For example, what is responsible for quantitative differences in product expression of the same compounds in closely related plants? Why are some pathways expressed in trace amounts? What is the genetic basis of controls of inducible versus constitutive pathways? Does alteration of promotor or regulator genes result in expression of hidden pathways? Can such genes or genes for single enzymes be inserted into unrelated plants, causing expression of entirely new pathways or compounds? How substrate-specific are the enzymes of secondary metabolism? How many of these enzymes exist as multiple isozymes with different catalytic properties? Why are some pathways formed by polymeric assembly whereas others are not? How widespread are non-enzymic (i.e. chemical) conversions, and what determines whether chemical conversions can occur?

7. IMPLICATIONS
(a) The cost of plant defence

Cost is inherently difficult to estimate because secondary metabolism is an integrated component of whole-plant function. Compounds may protect against abiotic stress or damage and may serve both primary metabolic and defensive roles (Chew & Rodman 1979). However, our theory suggests that it may not be possible to obtain realistic estimates of cost, irrespective of these confounding factors. A number of attempts have been made to estimate cost based on a single compound or class of compounds that are active against one or a few consumers (see, for example, Krischik & Denno (1983); Coley (1986); Baldwin *et al.* (1990)). Because this approach assumes that these compounds constitute the defence of the plant, it is likely to grossly over- or underestimate the real aggregate cost of defence, depending on whether these compounds are major or minor components of the entire chemical diversity with all its attendant metabolic pathways and controls. It is quite likely that a different estimate would be obtained if different compounds active against different consumers were chosen.

Calculations that do not make assumptions about which compounds are active, but are based on the quantities of components, are more likely to produce accurate estimates of cost. However, such estimates (see, for example, Gulmon & Mooney (1986)) are usually based on the major components and their attendant pathways, and will tend to underestimate cost by excluding the cost of the host of other components comprising chemical diversity. This would be particularly true if most plants had a much greater diversity than is currently reported. Studies on the chemical diversity of the British umbellifers (Jones & Lawton 1991) reveal that the diversity of biosynthetically distinct pathways of secondary metabolites, let alone chemical diversity *per se*, is strongly positively correlated with the intensity with which the plants have been studied. Because most plants are not well studied, most estimates of cost based on component concentrations will be underestimates.

Cost estimates are of interest because cost–benefit relations are of paramount importance in evolutionary theories (McKey 1979; Rhoades 1979). A central part of our theory is that mechanisms must have evolved that ensure generation and retention of chemical diversity. Retention of inactive compounds is one way of increasing chemical diversity and the probability of producing active compounds in the face of consumer adaptation. Yet traditional views of cost–benefit demand that such compounds be lost by selection. This problem may be resolved if selection has favoured plants possessing general metabolic traits for generating diversity, and if the cost of retaining low concentrations of large numbers of inactive compounds is sufficiently low to be selectively neutral, or such costs are exceeded by benefits incurred by the increased probability of producing active compounds. Interestingly, the immune system of animals is an analogous means of generating massive chemical diversity at minimal cost. However, our understanding of the genetic and biochemical mechanisms of plant defence lags well behind our knowledge of the animal immune system.

(b) The evolution of plant defence

Our ideas have a number of important ramifications to theories on the evolution of plant defence. First, the presence of a high diversity of mostly inactive compounds implies that the relation between the specific chemical composition of a plant and its overall degree of defence will be weak, at best. Hence, the relations between contemporary chemistry and past biotic interactions may be even weaker. Many, perhaps most of the chemicals may not be serving a specific role, and may never have served any role, other than their general contribution to the diversity necessary to increase the probability of having a few active compounds. Given sufficient diversity, the probability of isolating an active compound is actually quite high. It may therefore be erroneous to assume that this compound, or the class to which it belongs, specifically evolved as a consequence of selection by any particular organism(s). A compound found to be active against an organism in one plant species, and found to be widely distributed in related plants, does not necessarily imply that such compounds specifically evolved as defences and were followed by subsequent speciation. It may be more parsimonious to assume that as diversification will most likely proceed within a class (see §5*a*), there is a high probability of chemical similarity that may operate independent of biological activity.

Second, correlations between the quantity or quality of secondary metabolites and consumer abundance or damage are highly likely to result in spurious statistical relations, simply by chance. This is because at high diversity there are many degrees of freedom in the independent chemical variables.

Third, it may be erroneous to assume that compounds in plants are inactive solely because of consumer adaptation. Furthermore, even if we were able to show a specific metabolic adaptation in a consumer that was lacking in other related organisms, it cannot be assumed that the inactive compound serves no purpose. Such a compound may be inactive now but active in the past; may be active against some other organism; and, most importantly, may have been inactive in the past, but via mutation, produced an active compound.

Fourth, conclusions drawn from artificial selection experiments that show that concentrations of an active chemical can be increased, or that composition of a class of chemicals can be shifted towards greater activity may be misleading if they assume that natural selection will produce the same result. Without detailed knowledge of the controls over diversification, it may not be possible to guess the likely path of chemical evolution.

Last, and most important, the general metabolic traits that confer diversity may have been selected for very early in the evolution of plants, perhaps before the evolution of terrestrial plants. Microorganisms and algae show very similar traits, which either implies convergence or perhaps an ancestral characteristic that was selected for in the earliest organisms. However, once these traits were present, the 'natural screening process' of diversification and elimination may proceed in a manner that was more or less independent of specific biotic interactions, provided that there was always selection for well-defended plants. This does not mean that canalization of metabolism has not been influenced by specific biotic interactions, but it does imply that such canalizations may not be critical to the evolution of most well-defended plants.

Thus, our theory can be considered a null model for the evolution of plant defence. Tight or diffuse coevolution (Ehrlich & Raven 1964; Fox 1981; Berenbaum 1983; Futuyma 1983) or even sequential evolution (Jermy 1984) may not be necessary. We envision a world in which the probability of biological activity is low; in which some damage to plants is always inevitable; in which consumer organisms can always adapt, often rapidly; and in which the composition of the consumer community is constantly changing. At its simplest, the sole requirement for plant defence under these conditions may be the continual production of a diversity of secondary metabolites. The evolution of plant defence may therefore have proceeded independent of consumer adaptation, once these fundamental traits were in place.

We thank John Lawton, Centre for Population Biology, for helpful discussion; the British Ecological Society for a Travelling Fellowship (C. G. J.), the Royal Society, CIBA Foundation (C. G. J. and R. D. F.), and the Mary Flagler Cary Charitable Trust for support (C. G. J.). Contribution to the program of the Institute of Ecosystem Studies.

REFERENCES

Baldwin, I. T., Sims, C. L. & Kean, S. E. 1990 Reproductive consequences associated with inducible alkaloidal responses in wild tobacco. *Ecology* **70**, 252–262.

Berenbaum, M. R. 1983 Coumarins and caterpillars: a case for coevolution. *Evolution* **37**, 163–179.

Berenbaum, M. R. 1985 Brementown revisited: Interactions among allelochemicals in plants. *Rec. Adv. Phytochem.* **19**, 139–169.

Bernays, E. A. & Woodhead, S. 1982 Plant phenols utilized as nutrients by a phytophagous insect. *Science, Wash.* **216**, 201–203.

Brattsten, L. B. 1979 Biochemical defense mechanisms in herbivores against plant allelochemicals. In *Herbivores, their interaction with secondary plant metabolites* (ed. G. A. Rosenthal & D. H. Janzen), pp. 199–270. New York: Academic Press.

Bryant, J. P., Chapin, F. S. III & Klein, D. R. 1983 Carbon/nutrient balance of boreal plants in relation to vertebrate herbivory. *Oikos* **40**, 357–368.

Bu'lock, J. D. 1965 *The biosynthesis of natural products.* London: McGraw-Hill.

Cates, R. G. & Redak, R. A. 1988 Variation in the terpene chemistry of Douglas-fir and its relationship to western spruce budworm success. In *Chemical mediation of coevolution* (ed. K. C. Spencer), pp. 317–344. San Diego: Academic Press.

Chapman, R. F. & Blaney, W. M. 1979 How animals perceive secondary compounds. In *Herbivores, their interaction with secondary plant metabolites* (ed. G. A. Rosenthal & D. H. Janzen), pp. 161–198. New York: Academic Press.

Chew, F. S. & Rodman, J. E. 1979 Plant resources for chemical defense. In *Herbivores, their interaction with secondary plant metabolites* (ed. G. A. Rosenthal & D. H. Janzen), pp. 271–307. New York: Academic Press.

Coley, P. D. 1986 Costs and benefits of defense by tannins in a neotropical tree. *Oecologia* **70**, 238–241.

Coley, P. D., Bryant, J. P. & Chapin, F. S. III. 1985 Resource availability and plant anti-herbivore defense. *Science, Wash.* **230**, 895–899.

Ehrlich, P. & Raven, P. H. 1964 Butterflies and plants: a study in coevolution. *Evolution* **18**, 586–608.

Fox, L. R. 1981 Defense and dynamics in plant–herbivore systems. *Am. Zool.* **21**, 853–864.

Futuyma, D. J. 1983 Evolutionary interactions among herbivorous insects and plants. In *Coevolution* (ed. D. J. Futuyma & M. Slatkin), pp. 207–231. Massachusetts: Sinauer.

Geissbühler, H., Brenneisen, P. & Fischer, H.-P. 1982 Frontiers in crop production: chemical research objectives. *Science, Wash.* **217**, 505–510.

Gershenzon, J. & Croteau, R. 1991 Terpenoids. In *Herbivores, their interaction with secondary plant metabolites* (ed. G. A. Rosenthal and D. H. Janzen). New York: Academic Press. (In the press.)

Goodman, L. S. & Gilman, A. 1965 *The pharmacological basis of therapeutics.* New York: Macmillan.

Gould, F. 1988 Genetics of pairwise and multispecies plant-herbivore coevolution. In *Chemical mediation of coevolution* (ed. K. C. Spencer), pp. 13–55. San Diego: Academic Press.

Gulmon, S. L. & Mooney, H. A. 1986 Cost of defense on plant productivity. In *On the economy of plant form and function* (ed. T. J. Givnish), pp. 681–698. Cambridge University Press.

Halbrock, K. & Grisebach, H. 1975 Biosynthesis of flavanoids. In *The flavanoids* (ed. J. B. Harborne, T. J. Mabry & H. Mabry). London: Chapman & Hall.

Haldane, J. B. S. 1949 (Reprinted 1985) Disease and evolution. In *On being the right size and other essays* (ed. J. Maynard-Smith), pp. 183–187. Oxford University Press.

Harborne, J. B. 1977 Chemosystematics and coevolution. *Pure Appl. Chem.* **49**, 1403–1421.

Herbert, R. B. 1981 *The biosynthesis of secondary metabolites.* London: Chapman & Hall.

Horsfall, J. & Cowling, E. B. (eds) 1980 *Plant disease.* New York: Academic Press.

Iwasaki, S. & Nozoe, S. 1974 Mono- and sesquiterpenes. In *Natural products chemistry*, vol. 1 (ed. K. Nakanishi, T. Goto, S. Ito, S. Natori & S. Nozoe), pp. 39–183. Tokyo, New York: Kodansha-Academic Press.

Jacobson, M. & Crosby, D. G. (eds) 1971 *Naturally occurring insecticides.* New York: Dekker.

Jermy, T. 1984 Evolution of insect/host plant relationships. *Am. Nat.* **124**, 609–630.

Jones, C. G. 1983 Phytochemical variation, colonization and insect communities: the case of bracken fern (*Pteridium aquilinum*). In *Variable plants and herbivores in natural and managed systems* (ed. R. F. Denno & M. S. McClure), pp. 513–558. New York: Academic Press.

Jones, C. G. & Firn, R. D. 1978 The role of phytoecdysteroids in bracken fern, *Pteridium aquilinum* (L.) Kuhn, as a defense against phytophagous insect attack. *J. chem. Ecol.* **4**, 117–138.

Jones, C. G. & Firn, R. D. 1979 Resistance of *Pteridium aquilinum* to attack by non-adapted phytophagous insects. *Biochem. System Ecol.* **7**, 95–101.

Jones, C. G. & Lawton, J. H. 1991 Plant chemistry and insect species richness of British umbellifers. *J. Anim. Ecol.* **60**. (In the press.)

Krischik, V. A. & Denno, R. F. 1983 Individual population and geographic patterns in plant defense. In *Variable plants and herbivores in natural and managed systems* (ed. R. F. Denno & M. S. McClure), pp. 463–512. New York: Academic Press.

Luckner, M. 1972 *Secondary metabolism in plants and animals.* London: Chapman & Hall.

McKey, D. 1979 The distribution of secondary compounds within plants. In *Herbivores, their interaction with secondary plant metabolites* (ed. G. A. Rosenthal & D. H. Janzen), pp. 55–133. New York: Academic Press.

Mann, J. 1987 *Secondary metabolism.* Oxford: Clarendon Press.

Neish, A. C. 1964 Major pathways of biosynthesis of phenols. In *Biochemistry of phenolic compounds* (ed. J. B. Harborne), pp. 295–359. London: Academic Press.

Pimental, D. & Bellotti, A. C. 1976 Parasite–host population systems and genetic stability. *Am. Nat.* **110**, 877–888.

Rhoades, D. F. 1979 Evolution of plant chemical defenses against herbivores. In *Herbivores, their interaction with secondary plant metabolites* (ed. G. A. Rosenthal & D. H. Janzen), pp. 3–54. New York: Academic Press.

Rosenthal, G. A. & Janzen, D. H. (eds) 1979 *Herbivores, their interaction with secondary plant metabolites.* New York: Academic Press.

Strong, D. H., Lawton, J. H. & Southwood, T. R. E. 1984 *Insects on plants: community patterns and mechanisms.* Oxford: Blackwell Scientific Publications.

Suffness, M. & Douros, J. 1979 Drugs of plant origin. In *Methods in cancer research*, vol. XVIA (ed. V. T. DeVita & H. Busch), pp. 73–126. New York: Academic Press.

Suffness, M. & Douros, J. 1982 Current status of the NCI plant and animal product program. *J. Nat. Prod.* **45**, 1–14.

Tallamy, D. & Raupp, M. (eds) 1991 *Phytochemical induction by herbivores.* New York: J. Wiley.

Templeton, W. 1969 *An introduction to the chemistry of the terpenoids and steroids.* London: Butterworths.

Discussion

S. B. Malcolm (*Department of Biology, Imperial College, Silwood Park, Ascot, U.K.*). I think one can argue the converse of Dr Jones's suggestion that plants maintain a high diversity of potential chemical defences against herbivores because of the low probability that new chemicals will be biologically active. Instead it may be possible that low probability will select for plants to exploit single metabolic pathways that have a high probability of producing biologically active chemicals. That taxonomically similar groups of plants rely to a large extent on single kinds of chemical defences argues for such 'molecular parsimony' and not evolutionary 'expensive' diversity to counteract low probabilities. Diversity of chemical defences can be covered by a single molecular type. For example, the toxic cardenolides of foxgloves, oleander and milkweeds, in three different families, are steroids with the same kind of toxic activity. Variation of this single molecular pattern appears to account for most of the defensive needs of the plants. These include the full range of toxic/digestibility reducing and mobile/immobile defences suggested by apparency and resource availability hypotheses. The reason that this single molecular pattern can cover the full range of defensive needs is that cardenolide toxicity is targeted at disrupting a fundamental physiological attribute of all herbivores: Na/K ATPases. The use of coumarins by umbellifers, alkaloids by Solanaceae, glucosinolates by Cruciferae, and cardenolides by Asclepiadaceae argues for molecular parsimony and selection to increase the probability that new molecules will be biologically active.

C. G. Jones. Is there clear evidence that 'taxonomically similar groups of plants' do 'rely to a large extent on single kinds of chemical defence', as a means of solving the problem of an inherently low probability of activity? First, the diversity of secondary metabolic pathways is not known for most plants irrespective of whether or not such compounds are biologically active. Well-studied species or genera (e.g. bracken, British umbellifers) have a demonstrably high pathway diversity. Perhaps the perceived lack of pathway diversity is because phytochemists tend to focus on particular classes of compounds in order to identify new structural variants; and chemotaxonomists often use these same variants to discriminate plants within related taxa. Second, perceived 'molecular parsimony' of defence may be an artifact of the tendency for studies on plant resistance to focus on activity of one or a few compounds against one or two organisms – usually dominant herbivores or major pests – rather than the much larger number of chemicals, herbivores and pathogens that may be involved in these interactions. More intensive studies (e.g. on bracken) reveal that resistance to a community involves different compounds within and between pathways. Thus it can be argued that the notion that the umbellifers use only coumarins, the Solanaceae use only alkaloids; the Cruciferae use only glucosinolates; and the Asclepidaceae use only cardenolides as defences is a reflection of study biases and inadequate data, rather than real 'molecular parsimony'.

From our toxicological perspective it is not possible that a 'diversity of chemical defenses can be covered by a single molecular type'. Screening studies show substantial differences in biological activity of the same or different compounds within and between target organisms. This is owing to both variation in the mode of action (i.e. which binding site or compound–compound interaction is involved) and variation in affinity (i.e. degree of binding to a particular site. This is equivalent to the dose required to effectively block all the receptor sites of a given type). Structurally dissimilar compounds usually bind to different types of receptors. Structurally similar compounds with the same mode of action can vary by orders of magnitude in their affinity to the same receptor site. Variation in receptor site structure between target organisms results in orders of magnitude variation in the degree of binding of the same compound. Consequently, from our toxicological perspective, a 'single molecular type' does not produce a diversity of different types of biological activities. The example of the common mechanism of cardenolides against certain Lepidoptera via Na/K ATPase inhibition supports our argument. Here, structurally similar compounds are not diverse defences, because only a single mode of action is involved. Nevertheless, variation in affinity does result in a different dose being required before biological activity is observed.

From our toxicological perspective, we do not distinguish between compounds termed 'toxic' versus those termed 'digestibility reducing' properties as in apparency theory. This distinction arose from the observed net effects on herbivores; reduced survivorship = 'toxic' and reduced growth = 'digestibility reducing'. We consider these terms to be toxicologically misleading because: (i) they equate net effects on herbivores to mode of action, when binding sites have not been determined; (ii) they emphasize differences between these two types of net effects, when toxicological studies show that both reduced survivorship and reduced growth can be end results of many different modes of action; (iii) they classify compounds such as tannins into one functional class (of 'digestibility reducers') that may not be justifiable toxicologically and (iv) they confuse mode of action with affinity (i.e. low dose required = 'toxin'; high dose required = 'digestibility reducer'). Whereas low dose required means high affinity and high dose required means low affinity, for the same mode of action. Toxicologically speaking, compounds vary in both mode of action and affinity (dose response) and the combination of these two characteristics results in net effects on herbivores. The relations among modes of action, affinity and net effects such as survivorship or growth have yet to be established. The terms 'mobile/immobile' refer to biosynthetic turnover of compounds and not biological activity. Tannins may be 'immobile' and monoterpenes may (or may not) turnover (i.e. be 'mobile'). However, the relation between these biosynthetic characteristics and biological activity has not been clearly established.

The evolution of cellulose digestion in insects

MICHAEL M. MARTIN

Department of Biology, University of Michigan, Ann Arbor, Michigan 48109–1048, U.S.A.

SUMMARY

Despite the abundance and diversity of species that include living or dead plant tissue in their diets, the ability to digest cellulose is rare in insects and is restricted to a small number of orders and families. In this paper it is argued that cellulolytic capacity is uncommon in insects simply because it is a trait that is rarely advantageous to possess. Although there is a growing body of evidence for the occurrence of symbiont-independent cellulose digestion in cockroaches and in higher termites from the subfamily Nasutitermitinae, cellulose digestion in insects is usually mediated by microorganisms. It is proposed that non-cellulolytic omnivorous scavengers and detritivores may be preadapted to evolve symbiont-mediated cellulolytic mechanisms because of the prevalence of mutualistic associations between such species and the microorganisms that normally reside in their hindguts. A scenario is proposed for the evolution of symbiont-mediated cellulolytic capacity in roaches and lower termites. Finally, it is suggested that biochemical studies of insect cellulases might provide crucial insights that would greatly advance our understanding of the evolution of cellulose digestion in insects.

1. INTRODUCTION

Considering the number and diversity of insects that thrive on diets consisting largely of plant tissue, it is striking how few are able to exploit the nutritive value of cellulose, which is doubtless the most abundant form of non-fossilized carbon on earth. Cellulose digestion is a complex process, involving a suite of enzymes with diverse modes of action. Although the production of a complete cellulolytic system is common among bacteria and fungi, it is unusual among animals. Cellulose digestion in animals is most often mediated by symbiotic cellulolytic microorganisms. Thus, a discussion of the evolution of cellulolytic capacity in insects requires that two major questions be addressed: (i) Why is cellulose digestion so rare, and (ii) why is symbiont-dependent cellulose digestion more common than symbiont-independent cellulose digestion? In this paper I first discuss some general aspects of the biochemistry of cellulose digestion. I then discuss the occurrence and mechanisms of cellulose digestion in insects, and review evidence for symbiont-independent cellulolysis. Finally, I conclude by posing tentative answers to the two questions posed above, and by suggesting future directions for research in this area.

2. THE ENZYMOLOGY OF CELLULOSE DIGESTION

All cellulolytic systems examined to date consist of a mixture of enzymes with different activities and specificities (Coughlan & Ljungdahl 1988). Cellulose digestion has been most thoroughly studied in white rot fungi, where the cellulase system consists of a mixture of three classes of soluble extracellular enzymes: (i)

endo-β-1,4-glucanases, which randomly cleave β-1,4-glucosidic bonds along a cellulose chain; (ii) exo-β-1,4-glucanases, which cleave cellobiose units from the non-reducing end of a cellulose chain; and (iii) β-1,4-glucosidases, which hydrolyse cellobiose and water soluble cellodextrins (oligosaccharides) to glucose. Endoglucanases generally exhibit very low activity toward crystalline cellulose, but are quite active toward amorphous cellulose and water-soluble derivatives of cellulose, such as carboxymethylcellulose (CMC), whereas exoglucanases are more active toward crystalline cellulose and less active against CMC (Knowles *et al.* 1988). Endo- and exocellulases work synergistically in bringing about the hydrolysis of crystalline lignocellulose, and effective cellulolysis requires the presence of both classes of enzymes.

The digestibility of native cellulose is highly dependent upon its crystallinity and its association with other structural polymers, especially lignin (Wood & Saddler 1988). The cellulolytic systems of some organisms incorporate processes that effect the removal of lignin and the disruption of crystalline structure (amorphogenesis). For example, oxidative processes that cleave cellulose chains and destroy the orderly arrangement of cellulose molecules are believed to act as amorphogenizing systems in some brown rot fungi. In insects, the grinding action of the mandibles and the highly alkaline conditions that prevail in the midguts of some species might also serve to reduce the crystallinity of ingested cellulose.

The characterization of insect cellulases has been complicated by the presence of many enzymes in the gut fluids of cellulolytic species that are not insect enzymes at all, but rather are enzymes produced by cellulolytic microbial symbionts. Many insects, including species that are not able to digest cellulose, secrete endoglucanases and β-glucosidases analogous to those produced by white rot fungi (Martin 1983, 1987). These species lack the ability to digest cellulose because of the failure to produce exoglucanases or to have effective amorphogenizing systems that convert cellulose into a form that can be attacked by endoglucanases. The production by insects of enzymes active against crystalline cellulose, has only recently been shown (§5).

3. THE OCCURRENCE OF CELLULOSE DIGESTION IN INSECTS

Cellulose digestion has been shown in 78 species of insects from 20 families representing eight orders (table 1). Taxa in which cellulolytic capacity is common include the thysanuran family Lepismatidae (silverfish and firebrats), which has been elevated to ordinal status and renamed Zygentoma by Kristensen (1981), the Isoptera (termites), and the three coleopteran families Anobiidae (furniture beetles and death watch beetles), Buprestidae (metallic wood borers), and Cerambycidae (long-horned beetles). Cellulolytic capacity is also probably common in the orthopteran superfamily Blattoidea (roaches), the coleopteran family Scarabaeidae (scarab beetles), the dipteran family Tipulidae (crane flies), and the hymenopteran family Siricidae (woodwasps).

Termites are the most efficient cellulose-digesters, with assimilation efficiencies often approaching 99%. The xylophagous larvae of siricid woodwasps and anobiid, buprestid and cerambycid beetles are somewhat less efficient, with assimilation efficiencies in the range 12–68%. Cellulose digestion also occurs in a number of omnivorous species, such as silverfish,

Table 1. *The occurrence of cellulose digestion in insects*

The number in parentheses is the number of species in that family in which cellulolytic capacity has been shown. Species names can be found in Martin (1983, 1987)[a].

| order | order |
family	family
Thysanura	Coleoptera
Lepismatidae (5)	Scarabaeidae (2)
Orthoptera	Buprestidae (6)
Gryllidae (1)	Anobiidae (5)
Cryptocericidae (1)	Coccinellidae (1)
Blattidae (1)	Cerambycidae (29)
Blaberidae (1)	Curculionidae (1)
Isoptera	Trichoptera
Mastotermitidae (1)	Limnephilidae (1)
Kalotermitidae (2)	Diptera
Hodotermitidae (1)	Tipulidae (2)
Rhinotermitidae (6)	Hymenoptera
Termitidae (8)	Siricidae (3)
Plecoptera	
Pteronarcyidae (1)	

[a] Studies not referenced in Martin (1987): Lindsay (1940); Modder (1964); Zinkler (1983); Griffiths & Cheshire (1987); Zinkler & Götze (1987); Chararas & Noirot (1988); Hogan *et al.* (1988*a, b*); Kukor *et al.* (1988); Rouland *et al.* (1988); Scrivener *et al.* (1989).

Phil. Trans. R. Soc. Lond. B (1991)

firebrats, and roaches, where efficiencies range from 40–90%. Cellulose digestion is rare in detritus feeders, and digestive efficiency is usually low-to-moderate, ranging from 11–50%. Cellulose digestion in foliage feeders is very rare.

4. THE MECHANISMS OF CELLULOSE DIGESTION IN INSECTS

Four mechanisms have been proposed to account for cellulose digestion in insects: (i) exploitation of the cellulolytic capacity of protozoan symbionts residing in the hindgut; (ii) exploitation of the cellulolytic capacity of bacteria residing in the hindgut; (iii) reliance upon fungal cellulases, originating in the food, that remain active in the gut following ingestion; and (iv) secretion by the insect of a complete cellulase system. The multiplicity of mechanism implies that cellulolytic capacity has evolved independently in different groups, which explains the irregular distribution of the trait among highly divergent taxa.

The first mechanism, the exploitation of protozoan symbionts in the gut, was the first to be demonstrated. The dependence of the lower termites *Reticulitermes flavipes* (Rhinotermitidae) and *Kalotermes flavicollis* (Kalotermitidae) and the wood roach, *Cryptocercus punctulatus* (Cryptocercidae), on hindgut protozoa for cellulose digestion has been recognized since the classic investigations of Cleveland (1924). The symbionts are anaerobic protozoa from unique genera of flagellates restricted to the guts of the lower termites and wood roaches.

The second mechanism, the exploitation of cellulolytic bacterial symbionts in the gut, has been assumed to explain the ability of many higher termites (Termitidae) to digest cellulose, but convincing evidence in support of this assumption is still lacking (Breznak 1982; O'Brien & Slaytor 1982). Hindgut bacteria have been shown to contribute to cellulose digestion in the American cockroach, *Periplaneta americana* (Blattidae) (Bignell 1977; Cruden & Markovetz 1979) and the rhinocerus beetle, *Oryctes nasicornis* (Scarabaeidae) (Bayon 1981), and seem likely to participate in the larvae of crane flies (Tipulidae) (Sinsabaugh *et al.* 1985; Griffiths & Cheshire 1987).

The third mechanism, the use of ingested fungal cellulases, was discovered in my laboratory. Ingested fungal enzymes are essential participants in cellulose digestion in the fungus-growing termite *Macrotermes natalensis* (Termitidae: Macrotermitinae) (Martin & Martin 1978), larvae of the siricid woodwasp *Sirex cyaneus* (Siricidae) (Kukor & Martin 1983), and the larvae of five species of cerambycid beetles (Kukor & Martin 1986; Kukor *et al.* 1988). This mechanism has also been shown in other species of fungus-growing termites (Abo-Khatwa 1978; Rouland *et al.* 1988), and probably accounts for the limited cellulolytic capacity of stonefly nymphs and caddisfly larvae (Sinsabaugh *et al.* 1985).

The fourth mechanism, the production by the insect of all of the enzymes necessary for cellulose digestion, remains controversial, and is discussed in detail in the next section.

5. THE EVIDENCE FOR SYMBIONT-INDEPENDENT DIGESTION OF CELLULOSE IN INSECTS

Symbiont-independent cellulose digestion has been proposed to occur in cerambycid beetle larvae, in silverfish and firebrats, and in one roach and several termite species. In this section, I conclude that only in the cases of the roach and termite species is the evidence for symbiont-independent cellulose digestion strong.

(a) *Cerambycid beetle larvae*

The ability of the wood-boring larvae of cerambycid beetles to digest structural polysaccharides has been the subject of investigations dating back to the early 1900s. Cellulolytic enzymes have been detected in the gut fluids of many species. Influenced by the investigations of Cleveland (1924) on the role of hindgut protozoa in cellulose digestion in lower termites, early investigators attributed cellulose digestion in wood-feeding beetles to symbiotic yeasts housed in evaginations of the anterior-most segment of the midgut (Buchner 1928). During the 1930s, however, it was conclusively demonstrated that these yeasts do not contribute to cellulose digestion (Mansour & Mansour-Bek 1934), whereupon it was concluded that the beetles must produce their own cellulolytic enzymes. No evidence in support of that conclusion was presented, however, and other possibilities were not considered. It is now known that ingested fungal enzymes bring about cellulose digestion in some cerambycid species, and it is possible that hindgut bacteria may be involved in others, but there still has been no convincing evidence presented for the occurrence of symbiont-independent cellulose digestion in any anobiid, buprestid or cerambycid species. Cellulose digestion in wood-boring beetles deserves further study.

(b) *Silverfish and firebrats*

Cellulolytic capacity has been shown in five species in the thysanuran family Lepismatidae, but evidence concerning the agents of cellulose digestion is conflicting. Lindsay (1940) claimed that cellulose digestion in *Ctenolepisma longicaudata* occurs in the crop, mediated by aerobic bacteria. Lasker & Giese (1956), on the other hand, concluded that cellulose digestion in *C. lineata* occurs in the midgut, mediated by cellulases secreted by the silverfish. More recently, Zinkler & Götze (1987) have asserted that cellulose digestion in *Thermobia domestica* occurs in the crop, mediated by insect enzymes secreted in the midgut and refluxed forward into the crop.

Despite the unresolved confusion in the literature, the study of Lasker & Giese (1956) is regularly cited as a definitive demonstration that the silverfish *C. lineata* secretes a cellulase system capable of digesting native cellulose. In that study, the authors produced aposymbiotic (symbiont-free) animals by subjecting eggs to surface sterilization and raising the neonates on a diet of sterile rolled oats. They observed that the activity of the midgut fluid toward insoluble cellulose and the ability of the silverfish to degrade [^{14}C]-labelled cellulose to [^{14}C]-labelled carbon dioxide were undiminished in the aposymbiotic animals when compared with animals that had their normal complement of gut microbes, and conclude that cellulose digestion is brought about by enzymes produced by the insects.

Despite the impeccable design of this experiment, several methodological ambiguities, not evident in 1956, preclude so definitive a conclusion. First, there is no indication that the purity of the labelled cellulose was checked before feeding it to the silverfish. Commercial labelled cellulose is notorious for the presence of labelled non-cellulosic impurities. Indeed, it is common to find that as much as 10% of the counts in some preparations can be extracted with water and an additional 15–20% can be extracted with base. Thus, the possibility cannot be ruled out that an impurity, not cellulose, was the source of the labelled carbon dioxide in the respired gasses collected from the aposymbiotic nymphs. Second, in assaying cellulolytic activity in the gut fluid, the substrate was 'regenerated cellulose' prepared according to Trager (1932). The preparation of 'regenerated cellulose' results in significant loss of crystallinity, making it susceptible to degradation by endoglucanases alone. Consequently, finding undiminished activity toward 'regenerated cellulose' in the midguts of the aposymbiotic animals does not establish that they still possess a complete cellulolytic system or that they are still able to digest native cellulose. Of course, these criticisms do not necessarily mean that Lasker & Giese are wrong in their conclusion that symbiont-independent cellulose digestion occurs in *C. lineata*, but only that the evidence in favour of that conclusion is not compelling. This important study deserves replication. The order Thysanura (or Zygentoma) includes the most primitive known insects. A knowledge of the mechanism by which these insects digest cellulose is a critical piece of information in any analysis of the evolution of cellulolytic capacity in insects.

(c) *Termites and roaches*

Studies of the higher termites *Trinervitermes trinervoides* (Nasutitermitinae) (Potts & Hewitt 1973, 1974), *Nasutitermes walkeri* and *N. exitiosus* (Nasutitermitinae) (Schulz *et al.* 1986; Hogan *et al.* 1988*a*) and the Australian wood-eating roach *Panesthia cribrata* (Blaberidae) (Scrivener *et al.* 1989) provide the strongest evidence to date for symbiont-independent cellulose digestion. The first indication that these species might not be dependent upon microbial symbionts for cellulose digestion was the demonstration that enzymes with activity toward crystalline cellulose are secreted by the midgut epithelium. The discovery of such enzymes does not, of course, show that they are the sole agents of cellulose digestion *in vivo*. Cellulose digestion could still be primarily a symbiont-mediated process, with insect-derived enzymes serving only in a subsidiary role. Indeed, the production of cellulases with detectable activity toward crystalline cellulose has been shown in three additional termite species (*Mastotermes darwiniensis*, *Coptotermes lacteus*, *Macrotermes mülleri*) in which cellulose digestion most likely does involve microbial symbionts. In *M. darwiniensis* (Mastotermitidae) (Veivers *et al.* 1982) and *C. lacteus* (Rhinotermitidae) (Hogan *et al.* 1988*b*), both of which are lower termites, hindgut protozoa are the most likely major agents of cellulolysis, and in *M. mülleri* (Termitidae: Macrotermitinae) (Routland *et al.* 1988), cellulose digestion is mediated by a mixture of termite enzymes and ingested fungal enzymes.

A demonstration of symbiont-independent cellulose digestion requires that evidence be obtained which shows a lack of participation by gut protozoa and bacteria. The involvement of symbiotic hindgut protozoa in cellulose digestion in the termites *N. walkeri* and *N. exitiosus* and the roach *P. cribrata* was easily ruled out. Protozoa do not occur in the hindguts of higher termites, and occur only in small numbers in the roach. Furthermore, elimination of protozoa from the roaches' hindguts had no effect on gut cellulase levels. As evidence against the participation of hindgut bacteria in cellulose digestion in *N. exitiosus*, Hogan *et al.* (1988*a*) cite their inability to detect soluble cellulases in the hindgut and the failure of treatments designed to solubilize membrane-bound bacterial cellulases (e.g. sonication and incubation with lysozyme) to increase the level of hindgut cellulase. As evidence against the importance of bacterial cellulolysis in *P. cribrata*, Scrivener *et al.* (1989) cite the failure of ingested tetracycline (which reduces the total counts of hindgut bacteria by 85%) to reduce the viability of the roaches on a cellulose diet or to reduce gut cellulase activity. These are very important studies, and my only reluctance in accepting the conclusion of the authors that cellulose digestion in these species is symbiont-independent is my concern that the evidence, strong though it certainly is, is not quite compelling. Is it not still possible that cellulose digestion is mediated by fastidious anaerobes that are not affected by tetracycline and whose cellulosomes are not dislodged from the cell membrane by using the techniques employed? Perhaps an electron microscopic study of the type that revealed the mode of action of bacteria during the degradation of plant tissue in the rumen (Dinsdale *et al.* 1978) would provide the definitive evidence necessary to preclude the involvement of hindgut bacteria in the degradation of plant cell wall polysaccharides in these insects.

6. WHY IS CELLULOSE DIGESTION RARE IN INSECTS?

Cellulose digestion generates glucose, which is fermented to volatile fatty acids (VFAs) in some insects. Glucose and VFAs can be used to meet an organism's carbon or energy requirement. However, the dietary factors that usually limit growth and reproduction in insect herbivores are nitrogen or water, not carbon or energy (Slansky & Scriber 1985). Most herbivorous insects are able to meet their carbon requirements easily by digesting such ubiquitous plant constituents as sucrose and other oligosaccharides, starch, phospholipids, neutral lipids, and proteins (Applebaum 1985). The capacity to exploit the energy content of cellulose is unlikely to confer any particular benefit on an insect that can meet its carbon and energy requirements by exploiting more easily digested constituents of the diet.

It is possible that fitness-related processes other than growth and reproduction might be energy limited. For example, flight and stridulation have high energy requirements, and the intensity and duration of these activities can be limited by the rate at which metabolic energy is supplied. However, the process of cellulose digestion is too slow to support an activity requiring a rapid rate of energy production. Because they can be more rapidly mobilized and digested than cellulose, metabolic reserves, such as glycogen, trehalose, proline and fat, or ingested nutrients, such as nectar and honeydew, are the metabolic fuels of choice for the support of these activities.

Therefore, I propose a very simple answer to the question posed in the heading to this section. I suggest that cellulolytic capacity is uncommon in insects simply because it is a trait that is rarely advantageous to possess.

I propose further that even among cellulose-digesting species there are some that derive little or no benefit from this capacity. When fed labelled purified cellulose, nymphs of the stonefly *Pteronarcys proteus* (Plecoptera) and larvae of the caddisfly *Pycnopsyche luculenta* (Trichoptera) digest 11 and 12%, respectively, of the cellulose they ingest (Sinsabaugh *et al.* 1985). Stoneflies and caddisflies acquire their capacities to digest cellulose by the ingestion of microbial cellulases present in their normal diets of microbe-rich leaf litter. There is no evidence that the carbon derived from cellulose digestion makes a crucial contribution to the carbon budgets of these insects, or, indeed, that fitness is enhanced in any way by the acquisition of cellulolytic capacity. It is entirely possible that these species benefit from consuming a microbe-rich substrate, not because it provides a source of cellulolytic enzymes that enable them to assimilate a fraction of the carbon they ingest, but rather because it provides a rich source of nitrogen or micronutrients. For these species, cellulolytic capacity may be an inconsequential by-product of an association with microorganisms that is beneficial for some other reason. Non-adaptive microbe-mediated cellulose digestion in insects that are dependent upon microbial associates for other reasons provides a possible clue to the answer to the question posed in the heading to the next section.

7. WHY IS SYMBIONT-DEPENDENT CELLULOSE DIGESTION MORE COMMON THAN SYMBIONT-INDEPENDENT CELLULOSE DIGESTION? OR IS IT?

The prevalence of symbiont-dependent *vis-a-vis* symbiont-independent cellulose digestion could reflect either the ease with which symbiont-mediated processes evolve or the difficulty with which cellulase systems entirely of insect origin evolve. These two

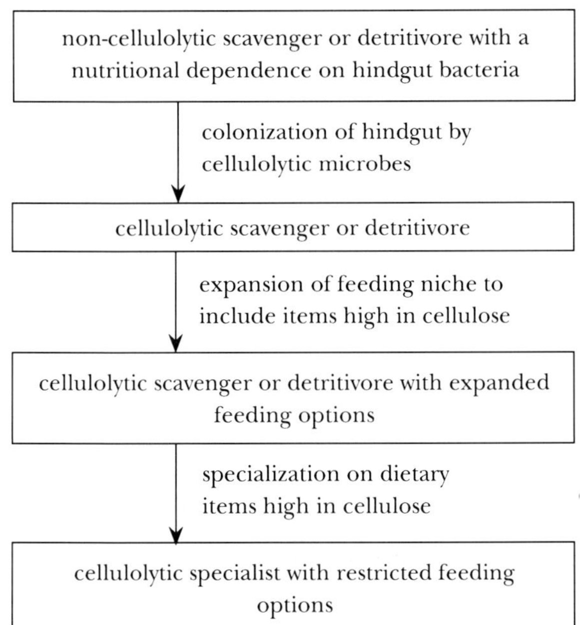

Figure 1. Scenario for the evolution of symbiont-mediated cellulose digestion in a scavenger or detritivore with a nutritional dependence on hindgut bacteria.

possibilities will be discussed separately. The possibility that the question is based upon an incorrect assumption will also be considered.

(a) What circumstances favour the evolution of symbiont-mediated cellulose digestion?

The hindguts of many omnivorous scavengers and detritivores, including roaches, crickets, scarab beetles and crane flies, provide conditions favourable for microbial growth and often harbour diverse and populous bacterial communities. Hindgut bacteria often make important contributions to their insect hosts. For example, in some species hindgut microbes supplement the nitrogen budget (by fixing nitrogen, conserving and upgrading uric acid nitrogen, or synthesizing essential amino acids) and provide critical micronutrients (vitamins and essential fatty acids) (Campbell 1989). The acquisition of cellulolytic capacity by an insect already dependent upon its hindgut bacteria for some other reason would only require that the hindgut become colonized by cellulolytic species. The resulting cellulolytic capacity might be entirely superfluous to the carbon budget of the host insect, but might also enable a change in the host's feeding niche to include dietary resources in which cellulose constituted the major carbon source. If a change in diet were followed by specialization on the new dietary resource, then cellulolytic capacity would no longer be superfluous, but would be indispensable.

This scenario (figure 1) can account for the pattern of cellulolytic capacity observed in roaches and lower termites, which are believed to have evolved from a roach-like ancestor. Most roaches are omnivorous scavengers. Their digestive enzymes include amylase, maltase, invertase, β-glucosidase, endocellulase, chitinase, esterase, lipase, and protease (Bignell 1981),

enzymes that enable them to exploit the easily digested carbon sources commonly present in many types of food. Roaches also possess an abundant hindgut flora, which is believed to contribute significantly to nutritional status by degrading and conserving uric acid nitrogen, supplying critical micronutrients, and possibly scavenging carbon by fermenting otherwise unabsorbable sugars to VFAs. Hindgut bacteria are also believed to be responsible for the cellulolytic capacities of roaches in the family Blattidae. The significance of cellulose digestion to the carbon budget of roaches on their natural diets is not known.

By contrast, in wood roaches (Cryptocercidae) and lower termites (Mastotermitidae, Kalotermitidae, Hodotermitidae, Serritermitidae, Rhinotermitidae), all of which are wood feeders, cellulose digestion plays a crucial role in the carbon economy. The hindguts of the lower termites and wood roaches, like those of omnivorous cockroaches, contain a diverse bacterial community, but in addition also include large populations of cellulolytic protozoa, which are the major agents of cellulose digestion. Deprived of their protozoan symbionts, wood roaches and lower termites are unable to survive for long on a diet of wood or cellulose (Cleveland 1924; Cleveland *et al.* 1934; Orlova 1974; Mauldin *et al.* 1972; Mauldin 1977). In the termite *Zootermoposis angusticollis* (Hodotermitidae), more than half of the carbon requirement is met with carbon derived from symbiont-mediated digestion (Hungate 1938).

The roach-like Australian termite, *Mastotermes darwiniensis*, is the most primitive termite. This species has characteristics expected of a transitional form between an omnivorous roach, in which cellulolytic capacity augments digestive scope but is not essential, and a xylophagous lower termite, in which cellulolytic capacity is essential and has been made more efficient by the colonization of the hindgut by cellulolytic protozoa. These termites not only attack timber, but also sugar cane, vegetables, and flour. Feeding workers a diet in which the carbon source is starch rather than cellulose results in the elimination from the hindgut of the four large protozoan symbionts normally found there and the loss of cellulase activity (Veivers *et al.* 1983). However, the termites produce an amylase and a maltase, and are able to survive indefinitely on a starch-containing diet. Cellulolytic capacity, although adaptive, is not indispensable, as this species can utilize sugar- and starch-containing foods, in addition to wood.

In summary, insects with diverse bacterial communities in their hindguts are pre-adapted to evolve symbiont-mediated celluloytic processes. Thus, the prevalence of symbiont-dependent cellulose digestion may simply reflect the prevalence of mutualistic associations between omnivorous scavengers and detritivores and their hindgut bacteria.

(b) What factors constrain the evolution of cellulolytic systems entirely of insect origin?

It is premature to suggest even a tentative answer to this question. Implicit in the question is the assumption that cellulolyti systems of insect origin are very rare, or at least are much less common that symbiont-mediated systems. That assumption may require revision. At the present time strong cases have been made for symbiont-independent cellulose digestion in only one species of cockroach and a few species of higher termites in the subfamily Nasutitermitinae. It is possible, however, that further research will reveal that cellulolytic systems entirely of insect origin are actually very common in roaches and higher termites. It is also possible that the early claim of symbiont-independent cellulose digestion in the Thysanura will be validated, and it is even possible that further studies of cellulose digestion in wood-feeding anobiid, buprestid and cerambycid beetle larvae will uncover additional examples. If symbiont-independent cellulose digestion proves to be more common than is currently assumed, the more appropriate question may be 'How do the circumstances that favour the evolution of cellulolytic systems entirely of insect origin differ from those that favour the evolution of symbiont-dependent cellulase systems?'.

On the other hand, there remains a small but finite possibility that the claims of symbiont-independent cellulose digestion in the roach *Panesthia cribata* and the nasute termites will not hold up. It is still possible that cellulose digestion in these species will ultimately prove to be mediated by gut bacteria. Furthermore, the early claim of symbiont-independent cellulose digestion in the Thysanura may not be confirmed, and additional research on cellulose digestion in wood-feeding beetle larvae may not provide any additional cases. The possibility that no insect digests cellulose without some microbial contribution cannot be totally discounted, in which case the appropriate question would be 'What has precluded the evolution of cellulolytic systems entirely of insect origin?'.

We would be well advised to know what the real question is before we hazard an answer.

8. FUTURE DIRECTIONS

Recent success in cloning cellulase genes (Béguin *et al.* 1987) has resulted in significant progress in elucidating structure–activity relations in the endo- and exoglucanases of bacteria (Kilburn *et al.* 1989) and fungi (Tomme *et al.* 1989). A common feature of these microbial cellulases is that they consist of two functionally independent domains, a catalytic domain that does not bind to cellulose and a cellulose-binding domain that is not enzymically active. An understanding of the functional roles of these two domains has led to important insights into the nature of the synergism that occurs between the exo- and endocellulases of microorganisms. Similar studies of the structure–function relations of insect cellulases might provide valuable clues to the factors that favour or constrain the evolution of both symbiont-dependent and symbiont-independent cellulolytic systems.

The study of cellulose digestion in insects has stimulated important research in insect physiology, and has played a seminal role in shaping the field of symbiosis research. This area remains an inviting one for investigators with interests that range from biochemistry and molecular biology to ecology and evolution.

REFERENCES

Abo-Khatwa, N. 1978 Cellulase of fungus-growing termites: a new hypothesis on its origin. *Experientia* **34**, 559–560.

Applebaum, S. W. 1985 Biochemistry of digestion. In *Comprehensive insect physiology biochemistry and pharmacology*, vol. 4 (ed. G. A. Kerkut & L. I. Gilbert), pp. 279–312. Oxford: Pergamon Press.

Bayon, C. 1981 Modifications ultrastructurales des parois végétales dans le tube digestif d'une larva xylophage *Oryctes nasicornis* (Coleoptera, Scarabaeidae): rôle des bactéries. *Can. J. Zool.* **59**, 2020–2029.

Béguin, P., Gilkes, N. R., Kilburn, D. G., Miller, R. C., O'Neill, G. P. & Warren, R. A. J. 1987 Cloning of cellulase genes. *CRC crit. Rev. Microbiol.* **6**, 129–162.

Bignell, D. E. 1977 An experimental study of cellulose and hemicellulose degradation in the alimentary canal of the American cockroach. *Can. J. Zool.* **55**, 579–589.

Bignell, D. E. 1981 Nutrition and digestion. In *The American cockroach* (ed. W. J. Bell & K. G. Adiyodi), pp. 57–86. London: Chapman & Hall.

Breznak, J. A. 1982 Intestinal microbiota of termites and other xylophagous insects. *A. Rev. Microbiol.* **36**, 323–343.

Buchner, P. 1928 *Holznahrung und Symbiose.* Berlin: Springer.

Campbell, B. C. 1989 On the role of microbial symbiotes in herbivorous insects. In *Insect-plant interactions*, vol. 1 (ed. E. A. Bernays), pp. 1–44. Boca Raton: CRC Press.

Chararas, C. & Noirot, C. 1988 Les osidases du termite *Nasutitermes lujae* (Termitidae). *Bull. Soc. Zool. Fr.* **113**, 175–180.

Cleveland, L. R. 1924 The physiology and symbiotic relationships between the intestinal protozoa of termites and their host, with special reference to *Reticulitermes flavipes* Kollar. *Biol. Bull.* **46**, 117–227.

Cleveland, L. R., Hall, S. R., Sanders, E. P. & Collier, J. 1934 The wood-feeding roach *Cryptocercus*, its protozoa, and the symbiosis between protozoa and roach. *Mem. Am. Acad. Arts Sci.* **17**, 185–342.

Coughlan, M. P. & Ljungdahl, L. G. 1988 Comparative biochemistry of fungal and bacterial cellulolytic enzyme systems. In *Biochemistry and genetics of cellulose degradation* (ed. J.-P. Aubert, P. Beguin & J. Millet), pp. 11–30. London: Academic Press.

Cruden, D. L. & Markovetz, A. J. 1979 Carboxymethylcellulose decomposition by intestinal bacteria of cockroaches. *Appl. Environ. Microbiol.* **38**, 369–372.

Dinsdale, D., Morris, E. J. & Bacon, J. S. D. 1978 Electron microscopy of the microbial populations present and their modes of attack on various cellulosic substrates undergoing digestion in the sheep rumen. *Appl. Environ. Microbiol.* **36**, 160–168.

Griffiths B. S. & Cheshire, M. V. 1987 Digestion and excretion of nitrogen and carbohydrate by the cranefly larva *Tipula paludosa* (Diptera: Tipulidae). *Insect Biochem.* **17**, 277–282.

Hogan, M., Veivers, P. C., Slaytor, M. & Czolij, R. T. 1988*a* The site of cellulose breakdown in higher termites (*Nasutitermes walkeri* and *Nasutitermes exitiosus*). *J. Insect Physiol.* **34**, 891–899.

Hogan, M., Schulz, M. W., Slaytor, M. Czolij, R. T. & O'Brien, R. W. 1988*b* Components of termite and protozoal cellulases from the lower termite, *Coptotermes lacteus* Froggatt. *Insect Biochem.* **18**, 45–51.

Hungate, R. E. 1938 Studies on the nutrition of *Zootermopsis*. II. The relative importance of the termite and the protozoa in wood digestion. *Ecology* **19**, 1–25.

Kilburn, D. G., Gilkes, N. R., Miller, R. C. & Warren, R. A. J. 1989 Cellulases of *Cellulomonas fimi*. In *Plant cell wall polymers* (ed. N. G. Lewis & M. G. Paice), pp. 587–596. Washington: American Chemical Society.

Knowles, J., Teeri, T. T., Lehtovaara, P., Penttilä, M. & Saloheimo, M. 1988 The use of gene technology to investigate fungal cellulolytic enzymes. In *Biochemistry and genetics of cellulose degradation* (ed. J.-P. Aubert, P. Beguin & J. Millet), pp. 151–169. London: Academic Press.

Kristensen, N. P. 1981 Phylogeny of insect orders. *A. Rev. Entomol.* **26**, 135–157.

Kukor, J. J. & Martin, M. M. 1983 Acquisition of digestive enzymes by siricid woodwasps from their fungal symbiont. *Science, Wash.* **220**, 1161–1163.

Kukor, J. J. & Martin, M. M. 1986 Cellulose digestion in *Monochamus marmorator* Kby. (Coleoptera: Cerambycidae): the role of acquired fungal enzymes. *J. chem. Ecol.* **12**, 1057–1070.

Kukor, J. J., Cowan, D. P. & Martin, M. M. 1988 The role of ingested fungal enzymes in cellulose digestion in larvae of cerambycid beetles. *Physiol. Zool.* **61**, 364–371.

Lasker, R. & Giese, A. C. 1956 Cellulose digestion in the silverfish *Ctenolepisma lineata*. *J. exp. Biol.* **33**, 542–553.

Lindsay, E. 1940 The biology of the silverfish, *Ctenolepisma longicaudata* Esch. with particular reference to its feeding habits. *Proc. R. Soc. Victoria (N.S.)* **52**, 35–83.

Mansour, K. & Mansour-Bek, J. J. 1934 The digestion of wood by insects and the supposed role of microorganisms. *Biol. Rev.* **9**, 363–382.

Martin, M. M. 1983 Cellulose digestion in insects. *Comp. Biochem. Physiol.* **75A**, 313–324.

Martin, M. M. 1987 *Invertebrate-microbial interactions: ingested fungal enzymes in arthropod biology.* Comstock Publishing Associates: Ithaca.

Martin, M. M. & Martin, J. S. 1978 Cellulose digestion in the midgut of the fungus-growing termite *Macrotermes natalensis*: the role of acquired digestive enzymes. *Science, Wash.* **199**, 1453–1455.

Mauldin, J. K. 1977 Cellulose catabolism and lipid synthesis by normally and abnormally faunated termites, *Reticulitermes flavipes*. *Insect Biochem.* **7**, 27–31.

Mauldin, J. K., Smythe, R. V. & Baxter, C. C. 1972 Cellulose catabolism and lipid synthesis by the subterranean termite, *Coptotermes formosanus*. *Insect Biochem.* **2**, 209–217.

Modder, W. W. D. 1964 The digestive enzymes in the alimentary system of *Acrotelsa collaris* (Thysanura: Lepismatidae). *Ceylon J. Sci. (Bio. Sci.)* **5**, 1–7.

O'Brien, R. W. & Slaytor, M. 1982 Role of microorganisms in the metabolism of termites. *Aust. J. biol. Sci.* **35**, 239–262.

Orlova, E. A. 1974 Influence of the intestinal symbiont complex on the intensity of food consumption and the longevity of the termites *Reticulitermes*. In *Termites: collected articles, Trans. Entomol. Div.*, no. 5 (ed. E. K. Zolotarev), pp. 165–180. Moscow: University Publishing House.

Potts, R. C. & Hewitt, P. H. 1973 The distribution of intestinal bacteria and cellulase activity in the harvester termite *Trinervitermes trinervoides* (Nasutitermitinae). *Insectes Sociaux* **20**, 215–220.

Potts, R. C. & Hewitt, P. H. 1974 Some properties and reaction characteristics of the partially purified cellulase from the termite *Trinervitermes trinervoides* (Nasutitermitinae). *Comp. Biochem. Physiol.* **47B**, 327–337.

Rouland, C., Civas, A., Renoux, J. & Patek, F. 1988 Synergistic activities of the enzymes involved in cellulose degradation, purified from *Macrotermes mülleri* and from its symbiotic fungus *Termitomyces* sp. *Comp. Biochem. Physiol.* **91B**, 459–465.

Schulz, M. W., Slaytor, M., Hogan, M. & O'Brien, R. W. 1986 Components of cellulase from the higher termite, *Nasutitermes walkeri*. *Insect Biochem.* **16**, 929–932.

Scrivener, A. M., Slaytor, M. & Rose, H. A. 1989 Symbiont-independent digestion of cellulose and starch in *Panesthia cribrata* Saussure, an Australian wood-eating cockroach. *J. Insect Physiol.* **35**, 935–941.

Sinsabaugh, R. L., Linkins, A. E. & Benfield, E. F. 1985 Cellulose digestion and assimilation by three leaf shredding aquatic insects. *Ecology* **66**, 1464–1471.

Slansky, F., Jr. & Scriber, J. M. 1985 Food consumption and utilization. In *Comprehensive insect physiology biochemistry and pharmacology*, vol. 4 (ed. G. A. Kerkut & L. I. Gilbert), pp. 87–164. Oxford: Pergamon Press.

Tomme, P., Heriban, V., Van Tilbeurgh, H. & Claeyssens, M. 1989 Specific assays, purification, and study of structure-activity relationships of cellulolytic enzymes. In *Plant cell wall polymers* (ed. N. G. Lewis & M. G. Paice), pp. 570–586. Washington: American Chemical Society.

Trager, W. 1932 A cellulase from the symbiotic intestinal flagellates of termites and of the roach, *Cryptocercus punctulatus*. *Biochem. J.* **26**, 1763–1771.

Veivers, P. C., Musca, A. M., O'Brien, R. W. & Slaytor, M. 1982 Digestive enzymes of the salivary glands and gut of *Mastotermes darwiniensis*. *Insect Biochem.* **12**, 35–40.

Veivers, P. C., O'Brien, R. W. & Slaytor, M. 1983 Selective defaunation of *Mastotermes darwiniensis* and its effect on cellulose and starch metabolism. *Insect Biochem.* **13**, 95–101.

Wood, T. M. & Saddler, J. N. 1988 Increasing the availability of cellulose in biomass. *Methods Enzymol.* **160**, 3–10.

Zinkler, D. 1983 Ecophysiological adaptations of litter-dwelling Collembola and tipulid larvae. In *New trends in soil biology* (ed. P. Lebrun, H. M. André, A. DeMedts, C. Grégoire-Wibo & G. Wauthy), pp. 335–343. Louvain-La-Neuve: Dieu Brichart.

Zinkler, D. & Götze, M. 1987 Cellulose digestion by the firebrat *Thermobia domestica*. *Comp. Biochem. Physiol.* **88B**, 661–666.

Discussion

C. G. JONES (*Institute of Ecosystem Studies, Millbrook, New York, U.S.A.*). Is there any evidence that cellulose digestion enhances extraction or utilization of nitrogen?

M. M. MARTIN. I'm not aware of any evidence that cellulose digestion enhances extraction or utilization of nitrogen, but it is certainly a plausible suggestion. There is a little bit of protein associated with plant cell walls, which might be made available if the cellulose were digested. I don't have any information about the nutritional quality of that protein, although my recollection is that it isn't very good. However, cell wall lysis, chemical or mechanical, does not appear to be a prerequisite for the utilization of cell contents by insect herbivores. Apparently the cell walls are, or become, very porous under conditions that prevail in the guts of caterpillars, which neither digest cellulose nor grind up their food into small particles, so that cell contents leak out. This is work done by Dr Ray Barbehenn while he was a student of Professor Liz Bernays'. A paper describing it is currently under review.

E. A. BERNAYS (*Department of Entomology, University of Arizona, U.S.A.*). Professor Martin's story is certainly convincing. I wondered also about the pointlessness of digesting cellulose when it is nitrogen that is limiting. I thought perhaps that by digesting cellulose, nitrogen would be made available. Being an unbalanced suite of amino acids the symbionts would have the value of upgrading this unsuitable mixture to one that would actually be useful.

M. M. MARTIN. I would generalize Professor Bernay's point a little further. I suspect that supporting a diverse and metabolically active gut flora is often the most beneficial consequence of cellulose digestion, whatever the contribution of the gut flora to the insect host might happen to be. Upgrading nitrogen quality is certainly one way that feeding glucose to gut bacteria might benefit a cellulose-digesting insect. Others might be vitamin synthesis, assimilation of sulphate into sulfur-containing amino acids, nitrogen fixation, maintenance of reducing conditions, or metabolism of allelochemicals.